Sensors 2021

Topical Conference at the 2021 AIChE Annual Meeting

Boston, Massachusetts, USA and Online
7 - 11 November and 15 - 19 November 2021

ISBN: 978-1-7138-5694-8

Printed from e-media with permission by:

Curran Associates, Inc.
57 Morehouse Lane
Red Hook, NY 12571

Printed with permission by Curran Associates, Inc. (2022)

For permission requests, please contact AIChE
at the address below.

AIChE
120 Wall Street, FL 23
New York, NY 10005-4020

Phone: (800) 242-4363
Fax: (203) 775-5177

www.aiche.org

Additional copies of this publication are available from:

Curran Associates, Inc.
57 Morehouse Lane
Red Hook, NY 12571 USA
Phone: 845-758-0400
Fax: 845-758-2633
Email: curran@proceedings.com
Web: www.proceedings.com

TABLE OF CONTENTS

632970 Invited Talk - Transmission Reduction Artificial Intelligence System (TRAIS)

Abstract:

The 2020 coronavirus (COVID-19) pandemic infected millions of people across the world and is responsible for more than millions of deaths also.[1] The disease is caused by the Severe Acute Respiratory Syndrome Coronavirus 2 (SARS-CoV-2). The mortality rate is nearly 4% for those between the ages of 60-64 years old and a striking 13.4% for those in the 80+ age group.[2] Clearly, SARS-CoV-2 has been a massive risk given the widespread transmission of the virus.

SARS-CoV-2 is a beta-coronavirus closely related to the previously endemic SARS-CoV-1 and MERS-CoV, life-threatening respiratory diseases leading to pulmonary failure in patients, which caused outbreaks in Guangdong, China and Saudi Arabia, respectively.[3, 4] To date, there is a plethora of evidence[5-10] demonstrating that SARS-CoV-2 transmission occurs fundamentally from the spread of viral pathogens in the infected host's respiratory system to other susceptible hosts in contact with droplets, aerosols and fomites.[11]

Ninety-nine percent of aerosols produced by humans, regardless of age, sex, weight and height are less than 10 mm.[7, 14] This is concerning since the smaller the aerosol, the longer it takes to settle and the greater the probability of being inhaled by other individuals. In a turbulent atmosphere, particles of 100 mm take 5.8 seconds to settle while 0.5 mm aerosols take 41 hours to settle.[15] If aerosols contain viable pathogens, they are a threat even after settling down as they can cause objects to serve as sources of contamination.[16] In case of SARS-CoV-2, the viruses can be viable on a surface for up to 3 days.[17] Researchers have found SARS-CoV-2[9] with virulent activity[7] in collected aerosol particles from 0.2 mm to 10 mm, which is a serious concern for the spread of the disease through AC systems.[9] *This groundbreaking data strongly indicates the need of SARS-CoV-2 detection in air as well as instantaneous mitigation of the air to safeguard the public and public spaces of the populations.*

In this invited talk we present a ***Transmission Reduction Artificial intelligence (AI) System (TRAIS)*** for sensing, and removal of airborne particles*[19]*. The system continuously monitors aerosol particle concentrations using a prototyped ***Airborne Particle Sensing Module*** for low (0.2 - 2.5 mm) and high (2.5 - 10.0 mm) size particles. A connected mobile device with an ***Artificial Intelligence (AI) App*** turns ON the room's ***Ventilation Actuator Module*** via external hardware that plugs into the ventilation control system for the building to carry outdoor fresh air into the room and transport the high aerosol particle air to a disinfection system with a filter. The system design conserves outstanding ventilation energy for aerosol mitigation at an estimated daily cost savings (e.g. 23-fold if system activates only once in 24 hours), allowing for data-driven approaches to reduce viral transmission in hospitals without significant increases in ventilation energy consumption.

The present talk focuses on efforts to develop the TRAIS, including the integration to its components:[19] ***Airborne Particle Sensing Module***, ***Artificial Intelligence (AI) App*** and ***Ventilation Actuator Module***. In addition, we show how TRAIS was used to develop additional aerosol mitigation systems to safeguard the exposure to aerosols of healthcare workers in a hospital setting.

References

[1] WHO: WHO Coronavirus Disease (COVID-19) Dashboard. World Health Organization 2020.

[2] Banerjee A, Pasea L, Harris S, Gonzalez-Izquierdo A, Torralbo A, Shallcross L, Noursadeghi M, Pillay D, Sebire N, Holmes C, Pagel C, Wong W-K, Langenberg C, Williams B, Denaxas S, Hemingway H:

Estimating excess 1-year mortality associated with the COVID-19 pandemic according to underlying conditions and age: a population-based cohort study. Lancet 2020, 395:1715-25.

[3] I A, Ysrafil: Severe Acute Respiratory Syndrome Coronavirus 2 (SARS-CoV-2): An overview of viral structure and host response. Diabetes Metab Syndr 2020, 14:407-12.

[4] Che XY, Hao W, Wang Y, Di B, Yin K, Xu YC, Feng CS, Wan ZY, Cheng VC, Yuen KY: Nucleocapsid protein as early diagnostic marker for SARS. Emerg Infect Dis 2004, 10:1947-9, doi:10.3201/eid1011.040516.

[5] Mizumoto K, Kagaya K, Zarebski A, Chowell G: Estimating the asymptomatic proportion of coronavirus disease 2019 (COVID-19) cases on board the Diamond Princess cruise ship, Yokohama, Japan. Euro Surveill 2020, 25:doi: 10.2807/1560-7917.ES.2020.25.10.2000180.

[6] Lines P: Grand Princess Updates. Available online at: https://wwwprincesscom/news/notices_and_advisories/notices/grand-princess-updateshtml (accessed Sep 2020) 2020.

[7] Santarpia JL, Rivera DN, Herrera V, Morwitzer MJ, Creager H, Santarpia GW, Crown KK, Brett-Major D, Schnaubelt E, Broadhurst MJ, Lawler JV, Reid SP, Lowe JJ: Aerosol and Surface Transmission Potential of SARS-CoV-2. medRxiv 2020, 2020.03.23.:https://doi.org/10.1101/2020.03.23.20039446.

[8] Li Y, Leung GM, Tang JW, Yang X, Chao CYH, Lin JZ, Lu JW, Nielsen PV, Niu J, Qian H, Sleigh AC, Su HJJ, Sundell J, Wong TW, Yuen PL: Role of ventilation in airborne transmission of infectious agents in the built environment – a multidisciplinary systematic review. Indoor Air, International Journal of Indoor Environment and Health 2007, 17:2-18.

[9] Liu Y, Ning Z, Chen Y, Guo M, Liu YL, Gali NK, Sun L, Duan YS, Cai J, Westerdahl D, Liu XJ, Xu K, Ho KF, Kan HD, Fu QY, Lan K: Aerodynamic analysis of SARS-CoV-2 in two Wuhan hospitals. Nature 2020.

[10] Tufekci Z: We Need to Talk About Ventilation. How is it that six months into a respiratory pandemic, we are still doing so little to mitigate airborne transmission? The Altantic 2020, July 30:https://www.theatlantic.com/health/archive/2020/07/why-arent-we-talking-more-about-airborne-transmission/614737/?utm_source=Global+Wellness+Institute&utm_campaign=2885d9ec32-EMAIL_CAMPAIGN_BRIEF_2020_8_12&utm_medium=email&utm_term=0_bbb41a322d-2885d9ec32-69722529.

[11] Galbadage T, Peterson BM, Gunasekera RS: Does COVID-19 spread thorugh droplets alone? Front Public Health 2020, 24 April:https://doi.org/10.3389/fpubh.2020.00163.

[12] Gameiro da Silva M: Research Gate. An analysis of the transmission modes of COVID-19 in light of the concepts of Indoor Air Quality, 2020, April:DOI: 10.13140/RG.2.2.28663.78240.

[13] Correia G, Rodrigues L, Gameiro da Silva M, Gonçalves T: Airborne route and bad use of ventilation systems as non-negligible factors in SARS-CoV-2 transmission. Med Hypotheses 2020, 141:doi:10.1016/j.mehy.2020.109781.

[14] Zayas G, Chiang MC, Wong E, MacDonald F, Lange CF, Senthilselvan A, King M: Cough aerosol in healthy participants: fundamental knowledge to optimize droplet-spread infectious respiratory disease management. Bmc Pulm Med 2012, 12.

[15] Tran K, Cimon K, Severn M, Pessoa-Silva CL, Conly J: Aerosol generating procedures and risk of transmission of acute respiratory infections to healthcare workers: a systematic review. PLoS One 2012:https://doi.org/10.1371/journal.pone.0035797.

[16] Baron P: Generation and behavior of airborne particles.

https://wwwcdcgov/niosh/topics/aerosols/pdfs/aerosol_101pdf.

[17] van Doremalen N, Bushmaker Y, Morris C, Holbrook MG, Gamble A, Williamson BN, Tamin A, Harcourt JL, Thornburg NJ, Gerber SI, Lloyd-Smith JO, de Wit E, Munster VJ: Aerosol and Surface Stability of SARS-CoV-2 as Compared with SARS-CoV-1. N Engl J Med 2020, 382:1564-7.

[18] Escombe A, Ticona E, Chávez-Pérez V, Espinoza M, Moore DAJ: Improving natural ventilation in hospital waiting and consulting rooms to reduce nosocomial tuberculosis transmission risk in a low resource setting. BMC Infect Dis 2019, 19:88. ublished 2019 Jan 25. doi:10.1186/s12879-019-3717-9.

[19] Forzani E, Patel B, McKay K, Pyznar G: System and method for mitigating airborne contamination in conditioned indoor environments. Provisional Patent 2020, Aug-19.

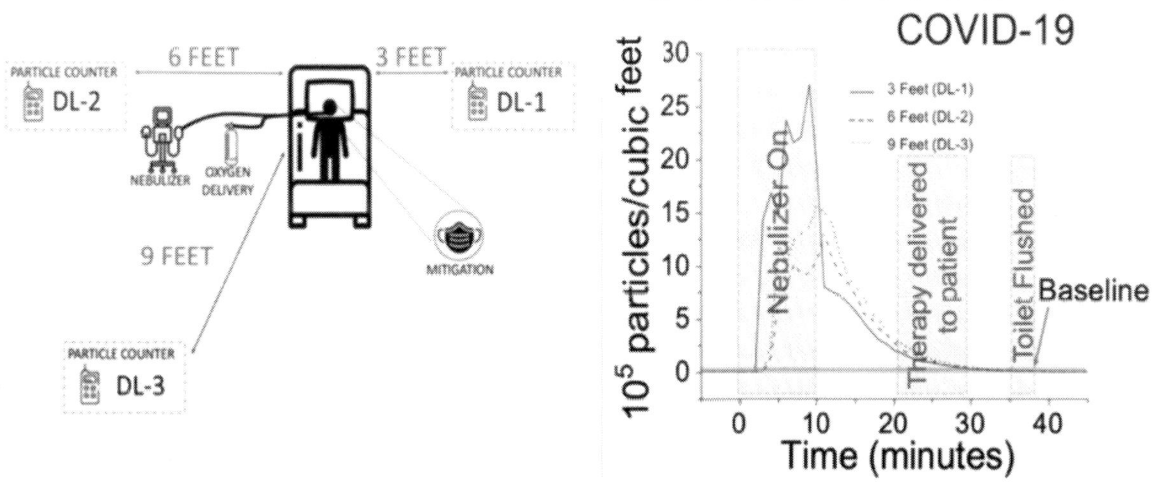

Fig. 1. Experimental setup and example of measurement from TRAIS for testing environments in COVID-19 patients at Mayo Clinic Hospital. The three aerosol particle sensors (DL1-3) from TRAIS were located at 3, 6, and 9 feet from the patient to determine the special-temporal distribution of aerosol particles in the room.

625153 Invited Talk - Carbon Nanotubes As Near-Infrared Fluorescence Biosensors of Intracellular Trafficking and Wound Monitoring

Abstract:

A challenge of particular interest in the fields of biosensing and bioimaging is the acquisition of real-time readouts of localized bioanalyte concentration through biomaterials and live tissue in a minimally-invasive fashion. The intrinsic fluorescence of single-walled carbon nanotubes (SWCNTs), which exhibits exceptional photostability, near-infrared (NIR) material-penetrating emission, and microenvironmental sensitivity, makes them ideal candidates for a variety of biomedical imaging and sensing applications. By functionalizing SWCNTs with appropriate biopolymers, we can simultaneously impart biocompatibility and sensitivity for certain biomolecules of interest. Using several novel spectroscopy, microscopy, and material processing approaches, this talk features our recent advances in the development of such engineered nanomaterials as NIR fluorescence biosensors. In particular, we have created *in vitro* sensors of endosomal trafficking in addition to the creation of microfiber-encapsulated SWCNTs for the real-time optical detection of oxidative stress in a wearable device.

628562 Mitigation of Positional-Dependent Limitations of Resonant Sensors and Applications Enabled Thereby

Abstract:

Resonant (LC) sensors are a class of wireless, passive (no battery) sensors; they consist of a simple conductive trace that resonate at a specific frequency which is interrogated with a proximal reader coil. The resonant frequency and magnitude of power reflection or transmission changes with variations in geometric features of the sensor or electrical properties of the environment. The sensor can be designed to transduce changes in temperature, biological growth, enzyme activity, motion, etc. in closed systems by monitoring the resonant features (scattering parameters via VNA). However, this interrogation through coupled electromagnetic fields of sensor and reader coil is also highly sensitive to the relative position between the sensor and reader coil. This limits resonant sensors to static applications where the relative position between the reader and sensor is fixed. We have developed two approaches to mitigate the positional-dependent signal and thereby extend the applications to which LC sensors can be used: 1) an array of sensors that allows for position determination and signal normalization and 2) an inductively-coupled extender (ICE) that is included between the reader coil and sensor to dramatically reduce the positional dependence. These two systems applied to changes in hydration including deep soil moisture sensing.

In the first approach, four square, planar resonators with unique frequency windows were used to form a 2 by 2 array for wireless position determination and normalization of position-dependent, embedded resonant sensors. First, a master table of S_{21} magnitude and phase data was collected at 8100 positions. Automated scripts extracted the characteristic magnitude and phase peaks and used cubic interpolation to expand the master table to 7,157,160 unique angle and coordinate positions. An unknown position is then determined by comparing its S_{21} measurements to this table. To further improve the position accuracy, multiple measurements are collected on linear flyby trajectories. The average and standard deviation of predicted position offset from true value using this method were 3.2 and 2.3 mm, respectively. To test normalization of a position dependent sensor, a spiral resonant sensor was placed underneath the square array. The sensor signal was modulated using varying amounts of water on the sensor surface. A corrected reading was determined using four different flyby trajectories using the position array data to adjust the signal based on position. We found that average errors of the normalized signals were between 0.04 to 0.15 MHz at lower water volume (0.5 mL) and -0.53 to -0.74 MHz at higher water volume (2.0 mL). In its current state, the positional array can be used for asset tracking or feedback control and the sensor normalization can be used to improve the measurement accuracy of embedded sensors. This technique can be further improved by collecting more accurate master calibration data using an automated system.

In the second approach we detail an inductively coupled extender (ICE) and resonant (LC) sensor to monitor soil moisture using a portable reader. Significant advances of this ICE-LC design are extending typical LC sensor read range over a meter and reducing positional alignment sensitivity between reader and sensor. An analytical model validates the working principle and feasibility of the ICE-LC system. Prototypes of the ICE-LC sensor were built and optimized in terms of sensitivity and power transfer (single and four turns for ICE top and bottom coils, respectively). Soil moisture tests validated the ICE-LC improvements on minimized positional alignment sensitivity and extended read range, transducing a decrease in resonant frequency with increasing soil moisture. When calibrating with existing wired approaches, the ICE-LC sensor had a reproducible, linear sensor gain of 4.52 %moisture content/MHz with an R^2 of 0.745 and RMSE of 2.41%. A smaller, planar form factor of the ICE-LC sensor was also tested and exhibited reduced positional alignment sensitivity between reader and sensor at shorter read ranges. This initial study demonstrates the feasibility of the ICE-LC resonant sensor as a cost-effective method to monitor soil moisture content throughout the growing season at many field locations.

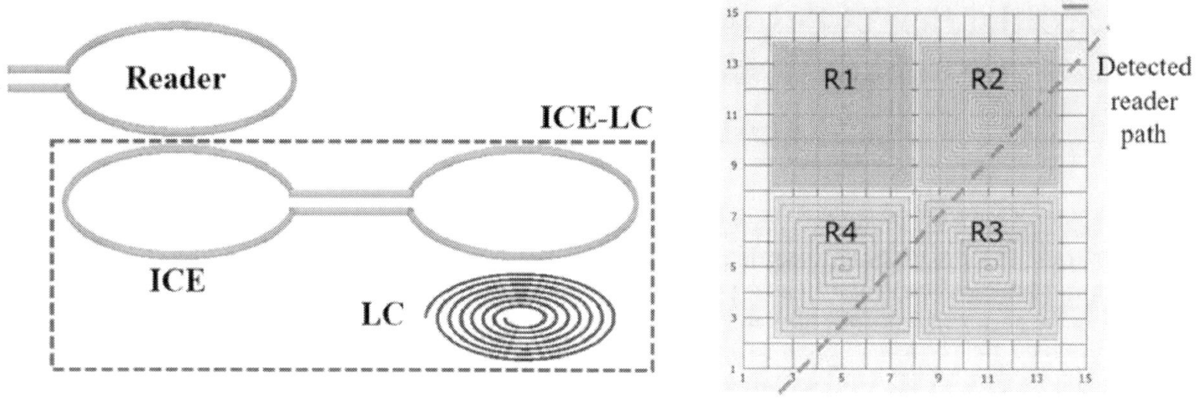

Left: An inductive-coupled extender (ICE) incorporated into the resonant (LC) sensor system to mitigate the position-dependent signal. Right: An array of four resonators used to provide the trajectory path and positions of the reader.

629720 On-Board Timing, Memory, and Sensing in Autonomous Cell-Sized Robots Enabled By a Simple Memristor-Based Circuit

Abstract:

Tetherless microscale robotic sensors which navigate enclosed spaces and remote locations are becoming a reality. While outstanding research has been done on robotic sensors actuated and supervised externally, progress is slow towards autonomous, intelligent cell-sized agents able to explore an unknown environment all by themselves: The seemingly straightforward path of miniaturizing macroscopic electronics poses an extremely high barrier-to-entry owing to the expertise required to design integrated circuit chips and their hundred-thousand-dollar cost of fabrication.

We demonstrate in this theoretical study that timing, memory, and sensing functionalities can be integrated within a sub-100μm robot with a lean, compact electrical circuit that only comprises simple resistive elements: just an array of memristors in parallel and (chemi)resistors in series. This dramatic reduction in complexity from a foundry-fabricated microchip additionally permits into the playing field a wide range of inexpensive technologies (printing, stamping, and colloidal assembly) as well as materials (polymers and nanomaterials).

Through simulations, we put our microrobotic sensor to the test for four progressively challenging autonomous tasks: (i) keeping track of the elapsed time since departure; (ii) detecting and timestamping a rare event; (iii) collecting and cataloging a series of time-indexed data; and (iv) accessing the stored data for scheduled and stimuli-responsive robotic actions. Through these tasks we establish the suitability of our design for real-world applications: leak detection inside of a remote pipeline, mapping the spatiotemporal analyte profile within an enclosed reactor or in the wild, and autonomous, analyte-responsive delivery of cargos such as insulin analogs. We expect the memristor-based design to become a universal platform for many upcoming autonomous microrobotic sensors in engineering, environmental monitoring, and healthcare as ubiquitous sensor nodes have become an increasingly important focus in the Internet-of-Things era.

630994 Detection of Benzene, Toluene, Ethylbenzene and Xylene in Air Using Micro-Preconcentrator/Solid Phase Micro-Extraction/Gas Chromatography Mass Spectrometry

Abstract:

The detection of volatile organic compounds (VOCs) is a thriving research area for many applications such as monitoring indoor air quality, environmental air, industrial safety, and health applications because of their potential adverse effects on human health and environment. Benzene, Toluene, Ethylbenzene, and Xylene (BTEX) are considered the most toxic VOCs due to their toxicity, carcinogenic nature, and inducing cardiovascular diseases at trace levels. Therefore, it is crucial to analyze these compounds accurately to monitor air quality. This work demonstrates the detection and quantification of trace BTEX in environmental air using a microfabricated preconcentrator combining with solid-phase micro-extraction/gas chromatography-mass spectrometry (SPME/GCMS) for concentration of these target analytes.

A micro-electro-mechanical system (MEMS) based preconcentrator was fabricated on a silicon wafer using deep reactive ion etching (DRIE). The inside cavity was packed with Carboxen-1000 adsorbent. 1L to 5L air samples were passed through micropreconcentrator where BTEX were adsorbed on the adsorbents at ambient temperature, then desorbed at 325° C and collected to a volume of 50 mL in a glass syringe using synthetic air. The adsorption and desorption flow parameters were optimized to 50 mL/min and 60 mL/min, respectively. Solid-phase micro-extraction for 15 minutes from the collected BTEX was performed using 75 μm Carboxen/Polydimethylsiloxane (CAR/PDMS) coated fiber. Then the fiber was inserted into GCMS to analyze compounds. Good linearity was observed for each standard analyte at a range of 1 ppb to 100 ppb (R^2 = 0.9988–0.9994) and the limits of detection (LOD) of the standard analytes achieved for benzene, toluene, ethylbenzene, m/p-xylenes, and o-xylene were 0.84, 0.85, 0.98, 0.82, 0.84 ppb, respectively. Three separate devices were compared by analyzing the same concentration of BTEX mixture and similar peak areas (relative standard deviation, % RSD < 8%) were observed, suggesting reproducibility among devices.

627661 Novel, Ultra-Small, Robust Electrochemical Sensor for in-Situ Detection of Cd(II) in the Environment

Abstract:

Heavy metal contamination is a rising global concern and causes a potent threat to humans and other biotas. Toxic metal species bioaccumulate through the food chain; thus, we are highly vulnerable to exposure to devastating health hazards. Most of the conventional metal-detecting analytical tools require lengthy sample pre-treatment methods, accordingly, alter metal speciation, one of the critical parameters for determining the toxicity of the metal species. Furthermore, these traditional techniques utilize less user-friendly bulky and expensive instruments, limiting real-time metal monitoring. Consequently, the development of a low-cost, portable, and robust sensor capable of providing accurate information on metal speciation will significantly aid in establishing metal mitigation systems efficiently.

In this study, we take advantage of ion transfer between two immiscible electrolyte solutions (ITIES) to develop a Cd(II) sensor. The fundamental mechanism of electrochemistry at ITIES, is that ion transfer is ruled by the ion's Gibbs energies of transfer for a specific aqueous-organic solvent system. Charged ions are transferred across the interface by imposing a potential difference (using a potentiostat) greater than the Gibbs energy for transfer between two phases and the resultant current can be measured as a function of applied potential. Electrochemistry at ITIES is less complicated since it doesn't involve electron transfer; hence more attractive over redox-based chemistry .A suitable ionophore is added to the organic phase to lower the energy requirement that needs to be applied via the external circuit when the Gibbs energy for ion transfer is too high (compared to that of background electrolyte solution). Moreover, this approach provides a greater selectivity for a particular target ion in the presence of a mixture of other ions.

Our electrode is a borosilicate glass electrode with an inner radius of ~300 nm. The nano-scale interface of our sensor follows hemispherical diffusion regime which allows us to have high mass transfer rate essential for fast kinetic measurements. An ionophore- 1-10 phenanthroline was used to facilitate the Cd(II) transfer across the nano-interface. We performed ITIES based cyclic voltammetry experiments with our nanosensor in various matrices, including simple electrolytes like KCl and complicated buffer solutions such as artificial seawater and artificial cerebellum fluid. We found out that our electrode shows excellent stability and capable of withstanding the complex matrices without fouling, an attractive feature of an exemplary sensor. We performed complexation studies with several ligands to show our sensor's capability to identify not just the unbound Cd(II) and the complexed, bound Cd(II). We tested our sensor with Cd(II) dissolved in a water sample collected from Indian River Lagoon, Melbourne, FL; thus, we showcase our sensor's power as an environmental monitoring tool. To the best of our knowledge, this is the first time reporting a glass electrode with a sub-nano-meter scale for Cd(II) detection in a natural environmental sample using ITIES. Our ultra-small electrode will enable us to study the kinetics of ion transfer across ITIES; thus, allowing us to modify the sensor to enhance the sensitivity and selectivity.

630439 Use of Dimensionality Reduction and Transfer Learning in Deep Reinforcement Learning Controller for Hydraulic Fracturing

Abstract:

Reinforcement Learning (RL) is an area of machine learning where an agent learns optimal control actions by directly interacting with the environment in order to maximize the cumulative sum of rewards. However, the application of RL to process control has been recent and limited [1]. There are multiple approaches to solving the RL problem in the literature but recent successes in RL has been due to combining concepts from deep learning with RL algorithms. The idea of leveraging deep neural networks (DNNs) as function approximators in RL has delivered tremendous success in video games like Atari games [2] etc. On the contrary, in process control, a Deep RL (DRL) controller was recently proposed to control discrete-time nonlinear processes [3]. The DRL controller is a model-free off-policy actor-critic algorithm in which learning is based on temporal difference (TD) error [1] and Deterministic Policy Gradient (DPG) theorem [4]. The actor and the critic are represented using DNNs so that the policy and the value functions can be generalized to the continuous state and action spaces. Ideas like replay memory, target networks, and gradient clipping were used in the DRL controller to make learning appropriate for process control applications. The DRL controller was able to track setpoints for single-input-single-output (SISO), multi-input-multi-output (MIMO), and a nonlinear system with external disturbances [3].

Hydraulic fracturing is the extraction of oil and gas from rocks with low porosity and permeability. In order to extract oil and gas from such rocks, an artificial medium of proppant (sand) is created within the fractures using controlled explosions followed by the pumping of fracturing fluids at high pressures. The efficiency of the extraction process is dependent on the final fracture geometry and proppant concentration along the fracture. In order to achieve the desired objectives with respect to these process variables, model predictive controllers (MPC) have been designed recently [5]. But these controllers (a) require an accurate model which is difficult to obtain in the case of hydraulic fracturing given its complexity, (b) involve exorbitant computational costs, and (c) require regular re-tuning of controller parameters. Therefore, in this work, we propose to design a model-free DRL controller for hydraulic fracturing. Despite its success, the DRL controller has a few limitations which include the requirement of long training times, and careful initialization of hyperparameters for fast convergence [6]. In order to overcome these challenges, we propose to implement transfer learning wherein the controller learns a suboptimal policy offline using a data-based reduced-order model (ROM) before learning online from the actual process. In this work, we used a high-fidelity model of hydraulic fracturing process as a virtual numerical experiment of the actual process. Additionally, we used principal component analysis (PCA) to reduce the dimension of the RL state before using it for learning. Consequently, the actor and the critic were trained in the reduced-PCA space. With these proposed steps included in the DRL controller training, it shows convergence to an optimal policy with the objective of obtaining uniform proppant concentration along the fracture.

Literature cited:

[1] Sutton, R.S., Barto, A.G. Reinforcement Learning: An Introduction. *Cambridge: MIT Press*, 1998.

[2] Mnih, V., Kavukcuoglu, K., Silver, D., Graves, A., Antonoglou, I., Wierstra, D., Riedmiller, M. Playing Atari with Deep Reinforcement Learning. *arXiv Preprint, arXiv:13125602*, 2013.

[3] Spielberg, S., Tulsyan, A., Lawrence, N.P., Loewen, P.D., Bhushan Gopaluni, R. Toward self-driving processes: A deep reinforcement learning approach to control. *AIChE J*, **65**(10), 2019.

[4] Silver, D., Lever, G., Heess, N., Degris, T., Wierstra, D., Riedmiller, M. Deterministic policy gradient algorithms. *In ICML*, 2014.

[5] Siddhamshetty, P., Yang, S., Kwon, J. S.-I. Modeling of hydraulic fracturing and designing of online pumping schedules to achieve uniform proppant concentration in conventional oil reservoirs. *Comput. Chem. Eng.*, **114**:306 – 317, 2018.

[6] Shin, J., Badgwell T.A., Liu, KH., Lee, J.H. Reinforcement Learning – Overview of recent progress and implications for process control. *Computer Aided Chemical Engineering*, **44**:71-85, 2018.

632115 Harnessing Cognitive AI Technology in Refining to Enhance Operator Decisions

Abstract:

This talk will explore a case study on how one of the largest refiners in the US is leading the way to implement Cognitive AI technologies that digitize refinery campaign processes from planning to operator action. We'll explain how this technology is helping operators to choose a judgment about what action to take at any given time, leading to better coordination and higher refinery margin through intermediate upgrades, product giveaway, and unit optimization.

Digital technologies and AI are becoming crucial for refiners to improve margins and reduce costs. There is a desire to move quickly so that you can be the company that achieves higher profitability before it becomes the standard. Yet we are observing inefficiency and resistance to digital adoption. Resistance and inefficiency persist at the business level from uncertainly in how to define the most impactful use case, and at a personnel level from a lack of trust in novel digital technologies. In this talk, we'll discuss how operations teams can begin building trust in AI technologies and share a roadmap of how a refinery launched a major initiative to optimize refinery processes and improve profitability by digitizing the refinery operating plan and leveraging best practices and knowledge to enhance operator decision making abilities.

Commercial optimization of a refinery is a unique exercise in many respects, providing both challenge and opportunity. These include crude slates that change according to market demand, variability in feedstock quality, and the continually changing operating capability of equipment from degradation and many other factors. What makes one refinery perform better than its competitors is often its ability to execute a set of objectives consistently while navigating through this uncertainty. According to Accenture, 61% of refiners believe process optimization technologies are a primary benefit driver to production planning and execution. Refinery campaigns are fundamental to improving margins but using a traditional problem-solving workflow between engineers and operators can be ineffective resulting in a very transactional execution process, relying heavily on individual interpretation of the guidance. The burden is on the operator to fully understand context, leading to inconsistencies over time and between different teams.

To close the talk, we'll highlight primary benefits from Cognitive AI technologies that are being used in refining today, including providing context to operators, quantifying decision confidence, and allowing for accelerated improvement across campaign plans, ultimately closing the profitability gap.

3 main takeaways:

1. How refiners can build trust in novel digital technologies including artificial intelligence and define the more impactful use cases for their refinery.
2. How one of the largest refineries in the US implemented Cognitive AI technologies to optimize refinery processes and improve profitability by digitizing refinery campaign processes from planning to action in operations.
3. New insights into how Cognitive AI technologies are helping operators to make a judgment about what action to take at any given time, leading to better coordination and helping refineries to better capture the full refinery margin.

629822 Regression Model for Tool Wear Monitoring in Precision Machining

Abstract:

In metal cutting operations, tool wear must be monitored either by operators or by a tool wear monitoring system to avoid possible damage to products and machine tools. In precision machining specifically, the naturally occurring and progressing flank wear leads to unwanted vibration, which deteriorates the surface finish and results in dimensional inaccuracy of the machining part. In the worst case, the cutting tool can break while it is engaged with the workpiece. With the ever-increasing demand for cost savings and improvements in quality, tool condition monitoring is becoming increasingly important in modern manufacturing [1].

Modern tool wear monitoring systems leverage data-driven models applying machine learning (ML) to predict tool wear and detect machine abnormalities [2]. Many studies have involved ML algorithms that use techniques such as K-nearest neighbors [3], support vector machine [4], and convolutional neural networks [5]. Although these methods have shown good performance, the fundamental methodology and complete toolchain that is suitable to predict tool wear still need to be investigated. Moreover, regression models that predict tool wear as a continuous process state/variable are seldom. The often-applied classification aims to predict discrete class labels, but tool wear is a continuously progressing system disturbance. Therefore, it is important to explore regression models to predict tool wear of continuous variables. The objective of this work is to propose a regression-based framework for predicting tool wear in CNC systems. We use support vector regression (SVR) as the ML algorithm, to provide insights into and quantify tool wear prediction in the machining process.

As shown in Fig.1, in the proposed framework, we collect, tabulate, and label audio and vibration data of the CNC process under varying tool wear and explore the performance of the proposed framework using regression. First, audio and vibration signals are converted to frequency domain through Fast Fourier Transform (FFT) to reduce the complexity of the audio and vibration time waveforms collected. Then, Principal Components Analysis (PCA) is used to eliminate features of small variance, increase the information content, reduce overfitting and computational effort. The selected features of audio and vibration signals are modeled and analyzed with an SVR model. The framework of Fig.1 was applied to a Mazak Variaxis 630-5X II T Mill/Turn Machine 40 hp 5-axis CNC machine. The experimental facility was instrumented to acquire process data for machine power, spindle power, spindle vibration, and to record audio/video and is MTConnect capable. Moreover, the machine was controlled to perform milling at various spindle speeds, widths of cut, depths of cut, and feed rates. Application of the framework of Fig. 1 for tool wear monitoring will be presented. In terms of the accuracy in tool wear prediction, it will be shown that the audio and vibration signals are sufficiently rich in information about tool wear so that a very accurate model (0.013 and 0.007 in RMSE, respectively) could be developed with those signals. Computation of the relative error between real tool wear measurement and predicted tool wear showed 1.54% error in audio signal-based regression and 3.1% error in vibration signal-based regression. Fusion of these signals further improved the accuracy of the regression model. In conclusion, the acoustic signals that often lead engineers and operators identify issues and anomalies in the operation of precision machining equipment were used successfully to learn ML models to predict tool wear in a very accurate manner. Extrapolation of these models in time to be useful in machine prognostics is currently underway and will be discussed in this presentation.

Acknowledgment: This material is based upon work supported by the U.S. Department of Energy's Office of Energy Efficiency and Renewable Energy (EERE) under the Advanced Manufacturing Office Award Number DE-EE0007613.

makes any warranty, express or implied, or assumes any legal liability or responsibility for the accuracy, completeness, or usefulness of any information, apparatus, product, or process disclosed, or represents that its use would not infringe privately owned rights. Reference herein to any specific commercial product, process, or service by trade name, trademark, manufacturer, or otherwise does not necessarily constitute or imply its endorsement, recommendation, or favoring by the United States Government or any agency thereof. The views and opinions of authors expressed herein do not necessarily state or reflect those of the United States Government or any agency thereof."

References

[1] H. Shao, H. L. Wang, and X. M. Zhao, "A cutting power model for tool wear monitoring in milling," *Int. J. Mach. Tools Manuf.*, vol. 44, no. 14, 2004, doi: 10.1016/j.ijmachtools.2004.05.003.

[2] Z. Li, R. Liu, and D. Wu, "Data-driven smart manufacturing: Tool wear monitoring with audio signals and machine learning," *J. Manuf. Process.*, vol. 48, 2019, doi: 10.1016/j.jmapro.2019.10.020.

[3] A. Jegorowa, J. Górski, J. Kurek, and M. Kruk, "Use of nearest neighbors (K-NN) algorithm in tool condition identification in the case of drilling in melamine faced particleboard," *Maderas Cienc. y Tecnol.*, vol. 22, no. 2, 2020, doi: 10.4067/S0718-221X2020005000205.

[4] D. Shi and N. N. Gindy, "Tool wear predictive model based on least squares support vector machines," *Mech. Syst. Signal Process.*, vol. 21, no. 4, 2007, doi: 10.1016/j.ymssp.2006.07.016.

[5] F. Jia, Y. Lei, J. Lin, X. Zhou, and N. Lu, "Deep neural networks: A promising tool for fault characteristic mining and intelligent diagnosis of rotating machinery with massive data," *Mech. Syst. Signal Process.*, vol. 72–73, 2016, doi: 10.1016/j.ymssp.2015.10.025.

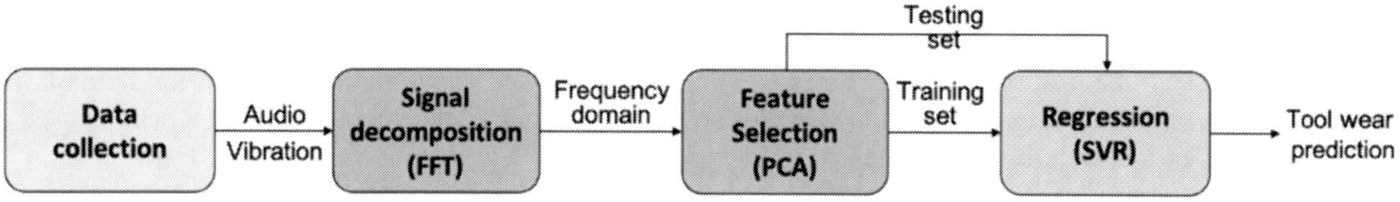

Figure 1. Flow diagram of a regression-based framework for tool wear prediction.

626215 Selection of Combined Index Weights to Optimize Anomaly Detection in Big Area Additive Manufacturing

Abstract:

As additive manufacturing (AM) becomes more prominent in our society, there are many questions about AM's repeatability and reliability. Today, AM is a process that involves constructing an object layer-by-layer; although this method allows the realization of complex computer-aided design (CAD) files into physical objects, a layer-by-layer process can also be more prone to error. Therefore, each layer has to be qualified so that the whole part can have a functional, rather than a purely aesthetic, purpose. Although AM has been used for several decades, most of the final part quality can be attributed to extensive operator knowledge and experience, rather than to the use of data-informed approaches. To assist operators, particularly those who may not be as knowledgeable about the process, and to monitor part quality, it is important to adapt data-driven approaches to warn operators of impending anomalies in the process. To achieve this, it is important to develop robust algorithms to interpret the vast quantities of collected data, especially when a priori knowledge of the data structure is nonexistent.

One traditional approach to monitor a manufacturing process is statistical process control (SPC). Before SPC is applied, a model that can adequately describe source of variation must be identified. Principal component analysis (PCA) [1] is one technique available to this end. The main idea is to describe normal variations by means of a number of principal components (PCs). This number (K) is smaller than the number of original variables (J). If there is no a priori knowledge about the data structure, it can be difficult to automatically calibrate a PCA model by selecting the correct number of PCs to retain. To overcome this issue, an existing proposed technique that involves the calculation of an ignorance score [2] to calibrate a probabilistic PCA model (PPCA) [3] is used in this study. An optimal model is selected by choosing the number of PCs for which the ignorance score achieves the minimal value.

After a model is selected using the ignorance score, SPC monitoring statistics such as Hotelling's T^2 statistic (H) and the squared prediction error (Q) can be used to detect anomalous data during process operation [4]. Anomalous data is identified when the H statistic and/or the Q statistic exceed their respective theoretical upper control limits (UCLs). While most applications make use of two charts, one for Q and one for H, a combined index () has been proposed more recently [5]. The combined index is a weighted sum of H and Q as seen in Eq. (1). The two weights, w_1 and w_2, are typically thought of as separate parameters [5]. So far, the weights of the combined index are calculated from UCLs for H and Q [5,6]. For example, the UCL of the H statistic relies on the assumption on how the H statistic is distributed (the chi-squared distribution was assumed in the original work [5]) and on selected confidence limits. Using UCLs as weights in the combined index is not motivated theoretically, only practically. Consequently, these UCL-based weights may render results obtained with the combined index difficult to translate from one application to the next. Today, there is no procedure available to set the weights (w_1 and w_2) so that optimal results can be assured, either theoretically or empirically. In this work, we present the first results of a systematic study where the chosen weights are varied deliberately in order to explore their effects on the combined index as a classifier during process monitoring.

Importantly, the combined index and its use for process monitoring is invariant to scale. Multiplying w_1 and w_2 by the same factor results in proportional change of the combined index and the upper control limit obtained for this combined index. To standardize our approach, we always chose the weights such that the sum of the eigenvalues of the empirical covariance matrix equals the following weighted sum of the same eigenvalues as seen in Eq. (2). The range of w_2 is chosen so that w_1 remains positive. Then, w_1 is calculated based on the above constraint.

Since the weights affect the estimated combined index, there likely exists an optimal set of relative weights that can optimize the performance of the combined index across numerous data types. The hypothesis is that the relative PPCA weights w_1=1 and w_2= (where) give the optimal anomaly detection performance. When these weights are applied, the combined index becomes equal to the squared Mahalanobis distance using the covariance matrix corresponding to the PPCA model [2,3] as seen in Eqs. (3-5). In essence, we are suggesting that the new optimal weights found in this study should be used consistently across literature to provide the best anomaly detection with the combined index.

The effect of the relative weights on the combined index is explored with simulated data sets [2]. Data types B1, B2, and C with noise to signal ratios up to 55% are used to generate simulated data sets with 1024 samples. The calculated combined indices are averaged over the number of samples. As seen in Figs. 1-2, the value of the combined index for a particular data set is a function of the selected weights. One should note that the vertical lines visually represent the locations of select weight ratios, such as the PPCA weights and UCL weights, where the chi-squared UCL has a specified confidence limit (i.e. 90%, 95%, and 99%). The simulated data sets allow the exploration of the effect of the noise to signal ratio and a nonoptimal number of retained PCs on the identification of the best performing relative weights. In order to compare the performance of each set of relative weights, the area under the curve (AUC) metric is used [7]. The AUC of the receiver operator characteristic curve (ROC) essentially describes the effectiveness of the combined index, given a set of relative weights, as a classifier to detect the true and false anomalies. The set of relative weights that lead to the maximum AUC (above 0.5 which is a random classifier) is the set of relative weights that produce the best classifier performance.

[1] Jolliffe, I.T. and Cadima, J., "Principal Component Analysis: A Review and Recent Developments." *Philosophical Transactions of the Royal Society A: Mathematical, Physical and Engineering Sciences*, 374(2065), 2016, pp. 20150202.

[2] Russo, S., Li, G., and Villez, K., "Automated Model Selection in Principal Component Analysis: A New Approach Based on the Cross-validated Ignorance Score," *Industrial & Engineering Chemistry Research*, 58(3), 2019, pp. 13448-13468.

[3] Tipping, M.E. and Bishop, C.M., "Probabilistic Principal Component Analysis," *Journal of the Royal Statistical Society: Series B (Statistical Methodology)*, 61(3), 1999, pp. 611-622.

[4] Montgomery, D., "Introduction to Statistical Quality Control", *Wiley*, 8th ed., 2019.

[5] Yue, H.H. and Qin, S.J., "Reconstruction-based Fault Identification Using a Combined Index," *Industrial & Engineering Chemistry Research*, 40(20), 2001, pp. 4403-4414.

[6] Jackson, J.E. and Mudholkar, G.S., "Control Procedures for Residuals Associated with Principal Component Analysis," *Technometrics*, 21(3), 1979, pp. 341-349.

[7] Fawcett, T., "An Introduction to ROC Analysis," *Pattern Recognition Letters*, 27(8), 2006, pp. 861-874.

$$\gamma = \frac{Q}{w_1} + \frac{H}{w_2} \quad (1)$$

where γ is the combined index, Q is the squared prediction error, H is Hotelling's T^2 statistic, and w_1 and w_2 are weights.

$$\sum_{j=1}^{J} \lambda_j = w_1 \sum_{j=1}^{K} \lambda_j + w_2 \sum_{j=K+1}^{J} \lambda_j = w_1 \sum_{j=1}^{K} \lambda_j + w_2\,(J-K) \quad (2)$$

where J represents the dimension of the data, K represents the number of retained PCs, $\sum_{j=1}^{K}\lambda_j$ is the sum of the retained eigenvalues, $\sum_{j=K+1}^{J}\lambda_j$ is the sum of the residual eigenvalues, and $\sum_{j=1}^{J}\lambda_j$ is the sum of all eigenvalues from the empirical covariance matrix.

$$\gamma_i = {y_i}^T(\Sigma)^{-1}y_i \quad (3)$$

$$\Sigma = w_1 \overline{P \Lambda P}^T + w_2 \underline{P P^T} = P \Lambda^* P^T \quad (4)$$

$$\Lambda^* = \begin{bmatrix} w_1 \overline{\Lambda} & 0 \\ 0 & w_2 I \end{bmatrix} \quad (5)$$

where γ_i represents the combined index of the ith row, y_i represents the ith row in the mean centered data matrix (n samples x J variables), Σ is the covariance matrix corresponding to the PPCA model, $\overline{P}$ are the retained PCs (J x K), $\overline{\Lambda}$ is a diagonal matrix of retained eigenvalues (K x K), $\underline{P}$ are the residual PCs ((J-K) x K), P are all of the PCs (J x J), and Λ^* is a diagonal matrix of all of the weighted eigenvalues (J x J).

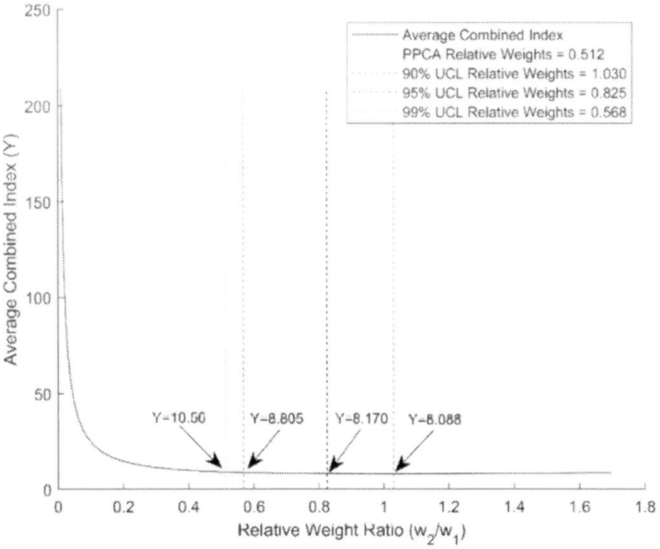

Figure 1. The combined index of data type B1.04 with a noise to signal ratio of 25% (calibrated with 3 PCs for a 9 dimensional data set). The vertical lines represent the locations of select relative weights. The theoretical UCL confidence level corresponds to the chi-squared control limit for the H statistic. The specified values on the plot represent the value of the combined index at the intersection of the selected relative weights indicated by the vertical lines.

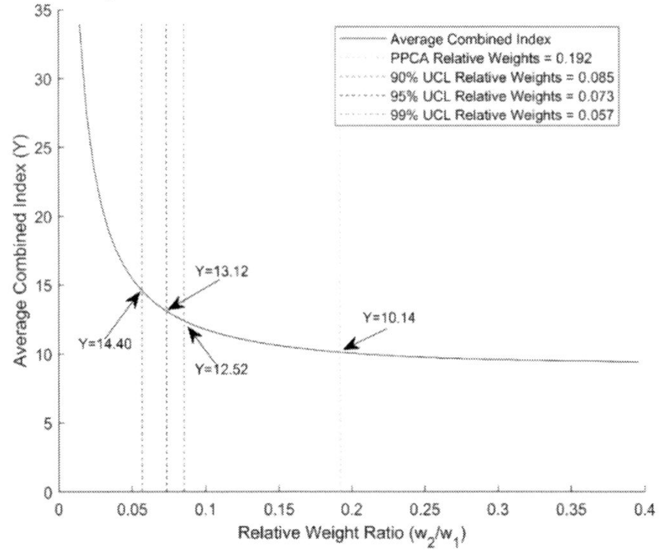

Figure 2. The combined index of data type C.02 with a noise to signal ratio of 25% (calibrated with 8 PCs for a 10 dimensional data set). The vertical lines represent the locations of select relative weights. The theoretical UCL confidence level corresponds to the chi-squared control limit for the H statistic. The specified values on the plot represent the value of the combined index at the intersection of the selected relative weights indicated by the vertical lines.

634105 Keynote Talk - Topological Data Analysis: Concepts, Computation, and Applications in Manufacturing

Abstract:

A key hypothesis that drives engineering practice is that data has structure. The dominant paradigms for describing such structure are statistics (e.g., moments, correlation functions) and signal processing (e.g., convolutional neural nets, Fourier series). Topological Data Analysis (TDA) is a field of mathematics that analyzes data from a fundamentally different perspective. TDA represents datasets as geometric objects and provides dimensionality reduction techniques that project such objects onto low-dimensional descriptors. The key properties of these descriptors (also known as topological features) are that they provide multiscale information and that they are stable under perturbations (e.g., noise, translation, and rotation). In this talk, we review key mathematical concepts and methods of TDA and present different applications in chemical engineering. Specifically, we discuss how to use TDA techniques to conduct real-time process monitoring, to design chemical sensors, and to quickly screen solvents for biomass processing.

634106 Keynote Talk - Platforms and Algorithms for Digitally-Enabled Next-Gen Manufacturing

Abstract:

Current initiatives calling for a significant increase in the degree of computing and communications networking technologies used to automate and optimize manufacturing can be traced to the 2011 – 2015 time period. The phrase "Industrie 4.0" was coined Wolfgang Wahlster and first used at the 2011 Hannover Fair in Germany (Kagermann, et al. (2011)). The phrase "Fourth Industrial Revolution" was defined in 2015 by Klaus Schwab (Schwab (2015)) and was used as the theme for the 2016 World Economic Forum Annual Meeting at Davos.

During the same time period, operating companies, notably ExxonMobil, issued a call for fundamental changes to the state-of-the-art of industrial control systems to address "pain points" including (1) a lack of interoperability between systems from different suppliers, (2) unacceptably high
costs for upgrades or replacements compared to other computing systems, and (3) restrictions on business value generation caused by barriers to insertion of third-party technology innovations (Bartusiak, et al. (2015)). This call to action resulted in the establishment of the Open Process Automation Forum of The Open Group in 2016 that is working to establish the industry standards for an open, secure, interoperable process automation architecture (https://www.opengroup.org/forum/open-process-automation-forum). The intent of the Open Process Automation Standard is to solve the identified industrial control system pain points and to enable realization of the vision of Industrie 4.0 and the Fourth Industrial Revolution for manufacturing.

Central to the Open Process Automation and Industrie 4.0 visions is a step-change increase in the pace of innovation with technologies including applied artificial intelligence, big data, and data analytics applied to manufacturing compared to the *status quo* with current industrial control system products that are of a closed, proprietary nature. Explicitly, one of the intentions of Open Process Automation is to increase opportunities for academic researchers and other technology providers to get their research transferred to industrial practice by reducing commercial barriers and the time it takes to deploy new technology.

In this talk, we will provide updates on the status of the Open Process Automation Standard (O-PAS), industrial prototypes of O-PAS based systems (Bartusiak, et al. (2021)), O-PAS conformance certification, and the state of Open Process Automation business ecosystem building. This will include guidance to academic researchers on how they can leverage the O-PAS Standard to facilitate the adoption of their research. We will comment on the relationships among Open Process Automation, Industrie 4.0, and similar initiatives -- such as the Clean Energy Smart Manufacturing Innovation Institute (CESMII) (Edgar and Pristikopoulis EN. (2018)), Manufacturing x Digital (MxD) (https://www.mxdusa.org), NAMUR Open Architecture (Klettner, et al. (2017)), and NAMUR Module Type Package (Bernshausen, et al. (2016)). Finally, we will comment on our outlook for new process systems engineering technologies -- including applied artificial intelligence, machine learning, big data, and data analytics -- to enable next-gen manufacturing efficiency.

References

Bartusiak RD, Bitar S, DeBari DL, Patin DL, Balsubramanian M, Foltz D, Caro RL, Brandl DL. (2015), "Functional Requirements for an Open Architecture Industrial Control System: Part 2 – What We Want". *ExxonMobil Research and Engineering Technical Report.*

Bartusiak RD, Bitar S, DeBari DL, Houk BG, Stevens D, Fitzpatrick B, Sloan P. (2021). "Open Process Automation: A standards-based, open, secure, interoperable process control architecture". *Proc. 2021 IFAC World Congress.*

Bernshausen J, Haller A, Holm, T, Hoernicke M, Obst, M, Ladiges J. (2016). "Namur Modul Type Package -- Definition". *atp-magazin* 58(1-2):72-81.

Edgar TF, Pistikopolous EN. (2018). "Clean Energy Smart Manufacturing Innovation Institute", Comp. Chem. Eng. 114:130-144.

Kagermann H, Lukas W-D, Wahlster W. (2011). "Industrie 4.0". VDI nachrichten. https://web.archive.org/web/20130304101009/http://www.vdi-nachrichten.com/artikel/Industrie-4-0-Mit-dem-Internet-der-Dinge-auf-dem-Weg-zur-4-industriellen-Revolution/52570/1

Klettner C, Tauchnitz T, Epple U, Northdurft L, Diedrich C, Schroeder T, Grossmann D, Banerjee S, Krauss M, Iatrou C, Urbas L. (2017). "Namur Open Architecture". *atp-magazin* 59(1-2):20-37.

Schwab K. (2015). "The Fourth Industrial Revolution". *Foreign Affairs.* https://www.weforum.org/agenda/2016/01/the-fourthindustrial-revolution-what-it-means-and-how-to-respond/

623852 Power and Lithography Free Highly Sensitive Digital Protein Detection Platform Using an Inexpensive Track-Etched Polycarbonate (PCTE) Membrane

Abstract:

Absolute and accurate quantification of several protein biomarkers present in a small volume of blood, cerebrospinal fluid, saliva, or urine can enable early disease diagnostics, prognosis and monitoring of further treatment in patients. Current state of the art technique for protein detection is Enzyme-Linked Immunosorbent Assay (ELISA) which suffers from low limit of detection (~nM-pM) beyond which several clinically relevant protein biomarkers (~pM-fM) of cancer, neurodegenerative, cardiovascular, inflammatory and autoimmune diseases remain undetected. In order to solve this issue, single-molecule arrays (Simoa) have been developed in which a protein molecule is sandwiched between an antibody coated magnetic beads and an enzyme conjugated detection antibody to form an immunocomplex. Subsequently, the immunocomplex is sampled into thousands of fL reaction chambers keeping the total protein concentration within the Poisson limit such that each bead contains either one or zero molecule. The fluorescence signal generated from the enzymatic reaction gives a digital readout providing an absolute quantification of the protein biomarker. However, Simoa are mostly performed in specially designed microwells which requires complex and expensive cleanroom based microfabrication techniques, costly and bulky experimental setup, fluidics control setups such as syringe pump, centrifuge, vacuum, or other specialized equipment's. Also, the total number of microwells present in the device which determines the sensors dynamic range is limited by the master mold which is cumbersome to tune. Moreover, these complex workflows require good amount of expertise and present a bottleneck for using them for point of care applications in resource limited areas where a use and throw sensor is preferable. In order to solve some of the abovementioned issues, we report a novel user friendly, power and clean room lithography free digital protein detection platform using an inexpensive, use and throw track-etched polycarbonate (PCTE) membrane.

Commercially available PCTE membranes are manufactured by ion irradiation in a nuclear reactor and have a high density (~10^6 pores/cm^2) of uniformly sized cylindrical through holes at various desired hole diameters with low protein binding properties. The through pores of a wetted membrane completely filled by capillary action can be easily converted into microwells when sealed on a sticky surface without generating any air bubbles which was confirmed from confocal microscopy; thus avoiding the use of vacuum pumps. Subsequently, when 2.8 μm sized magnetic beads are settled down by gravity on the 5 μm diameter membrane, less than 10% of the microwells are filled with magnetic beads. Single magnetic beads filling efficiency was further increased to more than 80% by keeping a magnet at the optimized distance at the bottom and mechanically shaking it. The microwells can then be easily isolated from each other by sealing with an oil phase making sure each microwell acts as an individual reaction chamber. A single enzymatic reaction produces a fluorescent product which is detected from standard fluorescent microscopy within 45 minutes as the reaction product is concentrated in a ~300 fL microwell reactor. Magnetic bead containing microwell that lights up will denote it's 'on' state and similarly, the ones that do not show any signal will denote 'off' state. The ratio of the enumerated 'on' and 'off' microwells using a custom written image analysis code correlated linearly with the protein concentration. Using this digital sensor, we achieve a dynamic range of pM-fM with a limit of detection of sub fM for biotin conjugated enzyme on streptavidin functionalized magnetic beads. This digital 'on' and 'off' readout format circumvents any bias that usually occurs in other analog sensors based on absolute intensity, current and voltage measurements.

The same device design was further developed to digitally detect specific extracellular vesicles (EVs) which are heterogeneous membrane-bound vesicles and are actively recognized as one of the most promising as well as stable biomarkers for liquid biopsy. Two color coded magnetic beads were functionalized with two different monoclonal antibodies respectively to capture EVs based on their surface protein from a 100ul sample. Then, the fluorescence signal generated from the enzymatic reaction is used to enumerate specific tumor-derived EVs based on the color of magnetic beads. Collectively, the inexpensive (< $0.1 per membrane) and user friendly PCTE membrane based digital ELISA can provide great platform for digital quantification of single protein biomarkers in a lithography and power free manner thus enabling point of care applications in laboratory resource constraints areas.

632256 Corona Phase Molecular Recognition of Interleukin 6 (IL-6) Using Near-Infrared Single Walled Carbon Nanotube (SWNT)

Abstract:

Interleukin-6 (IL-6) is a critical cytokine that regulates inflammation and plays an important role in numerous biochemical pathways. Currently, ELISA and western blots are the two primary IL-6 detection techniques. However, they require substantial cost, time, and could not offer timely analysis results. IL-6 biosensors capable of real-time, continuous measurements in *in vivo* environments, such as blood, interstitial fluid, or sweat, would greatly facilitate characterization of an organism's current inflammatory state, enabling new modalities of diagnosis and treatment of disease otherwise inaccessible. In this work, we generated synthetic recognition sites at the surface of a nanoparticle corona using Corona Phase Molecular Recognition (CoPhMoRe) for IL-6 detection. A library of 41 amphiphilic polymers varying in monomer composition and length were synthesized and interfaced with near-infrared fluorescent single walled carbon nanotube (SWNT) to create a sensor library. Two promising SWNT-polymer constructs have been discovered that show selective recognition of IL-6 among an analyte library of 19 biomolecules chosen for the members' structural diversity and *in vivo* relevance. Both sensors showed dose-dependent responses, and exhibited excellent stability with response from 10 ng/mL to 100 μg/mL. The IL-6 sensor is also deposited onto intravascular catheter and displays functionality in biological environment. The preliminary results demonstrate that synthetic molecular recognition strategies could enable a new class of continuous, *in vivo* monitoring of IL-6.

632840 Novel Probe for in-Situ Monitoring of Mitochondrial DNA Methylation Reveals Unsynchronized Nucleus and Mitochondria Epigenome Dynamics

Abstract:

Mitochondria, as the center of energy production, plays an important role in cell homeostasis by regulating cellular metabolism and mediating cellular response to stress. Epigenetic changes such as DNA and histone methylation have been increasingly recognized to play a significant role in homeostasis and stress response. The cross-talking between metabolome and epigenome has attracted significant attention in recent years but with a major focus on how metabolism contributes to epigenomic changes. Few studies have been focused on how epigenetic modifications may alter mitochondria composition and activity. The study of epigenetic regulation on mitochondria function is vastly limited by lack of epigenetic monitoring tools either on in nucleus or in mitochondria and lack of knowledge in epigenetic regulatory role on mitochondrial DNA (mtDNA). To fill this gap, we designed a novel probe targeting methylated CpGs of mtDNA and demonstrated the capability of our probe to reveal spatial distribution of methylated mtDNA and capture mtDNA methylation change at single cell level. We were also able to track single-cell mtDNA and nDNA methylation simultaneously and discovered unsynchronized dynamics from nucleus and mitochondria. Our tool offers a unique opportunity to understand epigenetic regulation of mtDNA and its dynamic response to microenvironment and cellular changes.

632966 Evaluation of Azure a Redox Dye As DNA Hybridization Indicator Using Electrochemical Methods

Abstract:

Over the last two decades, DNA-based electrochemical biosensors have captured an increasing attraction as base-pairing interactions between two complementary strands of nucleic acids offer high sensitivity and selectivity for biosensing applications. An electrochemical DNA biosensor can be designed either as label-based or label-free. The target DNA sequence or probe DNA receptor is labeled with redox-active molecules in label-based sensors while redox molecules such as intercalator dyes are used as DNA hybridization indicators in the label-free detection method. To date, various intercalator molecules have been used in biosensing applications such as methylene blue, ethidium bromide, doxorubicin, and meldola's blue. Here, we investigated the use of Azure A redox molecule, a phenothiazinium dye, in the intercalation-based DNA biosensor system developed to detect *the ctxA* gene of pathogenic *V. cholerae* bacteria.

For electrochemical sensing, the 5' of capture DNA probe was functionalized with biotin. High-affinity binding chemistry of streptavidin-biotin was used for attaching the biotinylated capture DNA probe onto the screen-printed carbon electrode surface. Selective hybridization between capture probe and complementary target DNA strands results in double stranded DNA sequences. The intercalation of redox dyes between the base pairs of DNA double helix allows long-range electron transfer between the Azure A and the carbon electrode surface, which results in a unique concentration dependent current change, distinct from the redox potential of Azure A, as measured by square wave voltammetry (SWV). Capture probe density, intercalator concentration, and intercalation duration were optimized. All SWV measurements were employed at 25 mV amplitude and 50 Hz frequency.

The results of this work suggest that Azure A may provide superior limits of detection for electrochemical intercalation-based biosensor systems.

632744 Detection of the Tuberculosis Volatile Organic Biomarker Methyl Nicotinate in Breath Using Electroactive Solutions (EAS)

Abstract:

Tuberculosis (TB) is a significant global health threat with the WHO reporting 1.4 million people deaths in 2019[1]. Breath biopsy methods present as a promising resource for TB diagnosis as it conjunctively negates the need for sputum collection that can cause many problems in young or ill patients and can provide results at the point-of-care. Methyl nicotinate (MN) has been identified by GC-MS as a specific biomarker of mycobacterium tuberculosis not readily found in the breath of healthy individuals[2], however routine testing through this method is inhibited by high operating costs and the undersupply of specially trained technicians. This semi-volatile biomarker has high solubility in water compared to other compounds found in breath. When passing breath into the EAS, it creates a well-filtered, concentrated sample ready for analysis electrochemically.

Our lab's electrochemical breath biopsy method utilizes a cobalt Engineered Electroactive Solution (EAS) to analyze samples on inexpensive, commercially available screen-printed electrodes via square wave voltammetry (SWV). The change in signal of a transition metal of known electrochemical activity is observed when the solution is exposed to an analyte of interest. These results show cobalt EAS to be both sensitive and specific to methyl nicotinate, responding to biomarker levels found in patient breath, with a limit of detention of approximately .01 mM MN. A unique peak for MN is observed with peak heights that proportionately increase with biomarker concentration. Furthermore, the use of cobalt EAS for detection of MN in mimicked patient breath samples shows that the diffusive gas to liquid transfer process adequately purified samples of potential confounders, resulting in negligible interference in detection.

[1] World Health Organization, "Status of the TB epidemic," 2019.

[2] Syhre, M. and Chambers, S.T., 2008. The scent of Mycobacterium tuberculosis. *Tuberculosis*, *88*(4), pp.317-323.

624217 Conformable Resonant Sensors for Contact-Free Sweat Analysis through Clothing and Protective Gear

Abstract:

Maintaining proper hydration is important to overall health and especially in work situations which require periods of high exertion. Monitoring sweat loss can help determine hydration status of individuals working in harsh conditions (high temperature, high humidity), which is especially relevant to those who wear thick personal protective equipment (PPE). This study presents a wireless, passive, skin-conformable sweat sensor sticker that can be worn under clothing and interrogated through thick PPE to simultaneously measure sweat loss volume and sweat conductivity. The sticker consists of a laser-ablated, microfluidic channel and a resonant sensor transducer. The resonant sensor is wirelessly read with a handheld vector network analyzer coupled to two, co-planar, interrogation antennas that measure the transmission loss. A sweat proxy is used to fill the channels and it is determined that the sensor can orthogonally determine the sweat conductivity and volume filled in the channel via peak transmission loss magnitude and frequency, respectively. Signal offset can be calibrated for by linear relationship between separation distance between reader and sensor by measuring the PPE thickness. A model is used to calibrate the sensor response to sweat conductivity and sweat loss with an R^2 of 0.964 and 0.934 respectively. A four-person study was performed to determine level of sensor variance caused by local tissue dielectric heterogeneity and sensor-reader orientation which has led to different resonator designs to diminish the effect of tissue dielectric and reader/sensor orientational sensitivity. This work provides a wearable sensor platform for human health monitoring which is passive (no on-board power), wireless, and can be interrogated through PPE; progress on extending this to skin temperature monitoring will be shared. It has potential applications in areas requiring thick PPE including in chemical industry, emergency response personnel, and military as well as competitive athletics (e.g., American football).

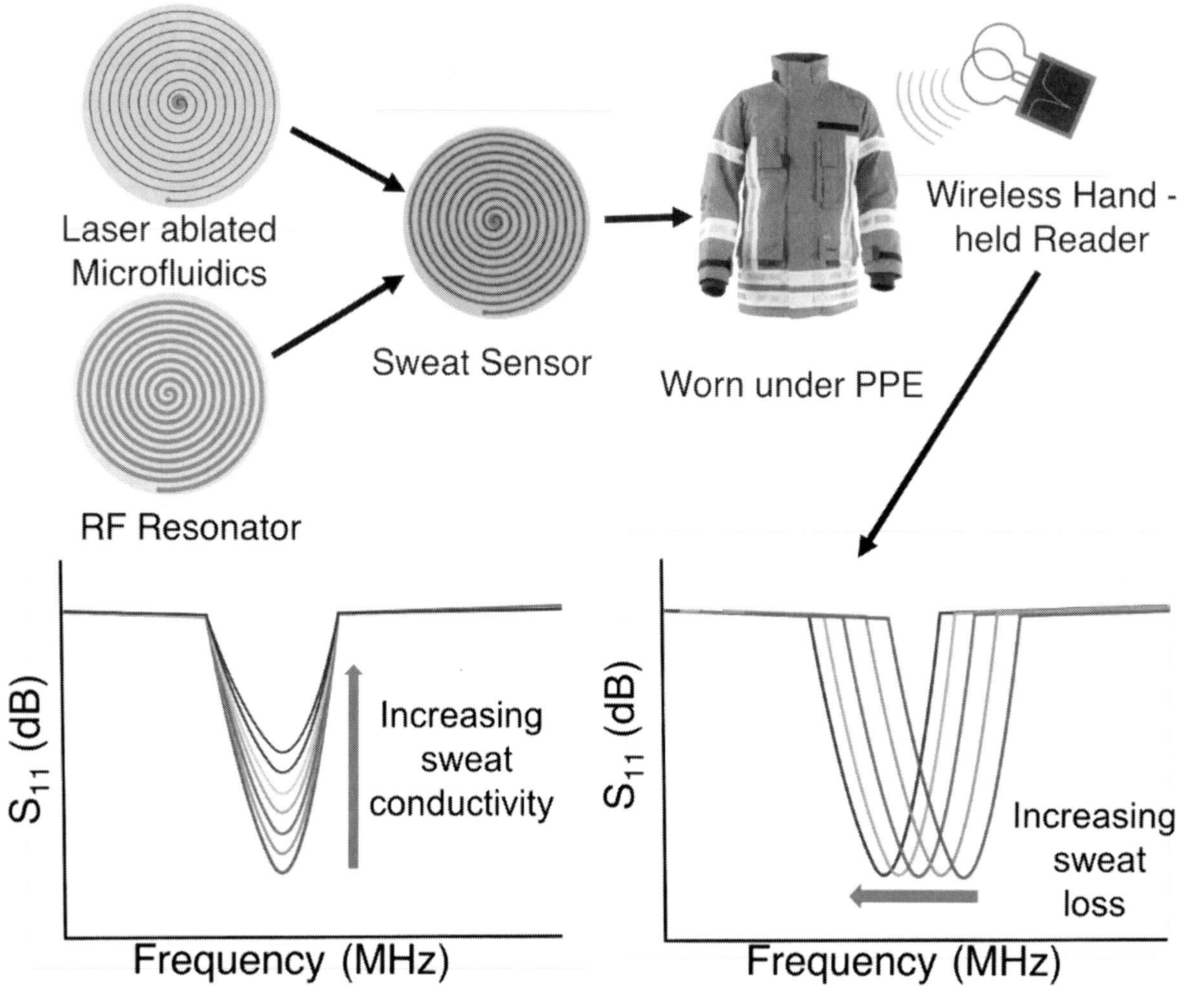
Laser ablated
Microfluidics
RF Resonator
Sweat Sensor
Worn under PPE
Wireless Hand -
held Reader
S_{11} (dB)
Increasing
sweat
conductivity
Frequency (MHz)
S_{11} (dB)
Increasing
sweat
loss
Frequency (MHz)

624738 Low-Cost Paper-Based Sensors for Ultra-Fast Room-Temperature Hydrogen Detection Using Palladium Alloy Nanowires

Abstract:

Use of hydrogen as a carbon-free energy carrier, particularly for use in fuel cell vehicles, is being widely explored and hydrogen-powered vehicles are beginning to enter the market. In this context, low-cost, low-power, widely-deployable hydrogen sensors are essential for safe operations, as H_2 forms explosive mixtures in air at concentrations as low as 4%. Paper-based sensors are of increasing interest for sensing applications due to their inherent advantages such as low-cost, portability, flexibility, and their environment-friendly degradable substrate. Paper-based sensors are effective for gas sensing applications due to their porous nature which enables more contact between the analyte and the sensing material, as well as their stability in the presence of most gaseous analytes. Penta-twinned palladium nanowires (PdNWs) are an excellent choice for sensitive H_2 gas-sensing applications due to their high aspect-ratio and electrical conductivity. Hydrogen adsorbs on Pd at ambient conditions, dissociates, and dissolves in the bulk to form palladium hydrides (PdH_x) with much lower conductivity than Pd. Thus, H_2 is detected based on a change in resistance. Here, we synthesized thin penta-twinned PdNWs in high yield using a hydrothermal method with polyvinylpyrrolidone (PVP) as a low-cost, high-molecular-weight surfactant to control growth and fabricated sensors by simple drop-casting. PdNWs drop-cast on an interdigitated electrode (IDE) on a silicon wafer gave a response of 0.3% to 1% H_2 in air, with response and recovery times of 12 s and 20 s, respectively. However, the PdNWs gave a negative response (decreased resistance) at lower concentrations due to the formation of new conduction channels between PdNWs upon volume expansion associated with hydrogen absorption and hydride formation. This has been widely observed in sensors based on Pd nanostructures, and results from poor contact between Pd nanostructures within the device. PdNWs on a paper substrate with hand-painted silver contacts had a more than 10-fold increase in response to 1% H_2, compared with those on an IDE, with improved response and recovery times of 10 s each. However, these sensors still had a negative response at lower concentrations of hydrogen. Such reverse sensing phenomenon is problematic for reliable hydrogen sensing over a broad range of concentrations. We observed that ozone treatment, by exposing the PdNWs-paper sensor to UV-O lights overnight, eliminates the PVP used in the synthesis of PdNWs and improves contact between the PdNWs, which eliminates the reverse sensing phenomenon. The UV-O-treated PdNWs sensor gave a response of 5% to 1% H_2 in air, with response and recovery times of 15 seconds. With elimination of the negative response at a lower concentration, we could easily detect as little as 100 ppm H_2 in air. Finally, upon coating the PdNWs with a small amount of Pt (<5% by weight), the response and recovery time improved dramatically to 5 s to 1% H_2. We note that the PdNW or PdPtNW loading per device is roughly 0.1 mg, corresponding to less than $0.01 of metal value. Hence, for the first time, we report a low-cost solution-processed palladium alloy (PdPt) nanowire-based paper sensor with ultra-fast response and recovery time of 5 seconds to 1% H_2 in air. These operate at room temperature, with minimal power requirements, and could therefore be widely deployed in support of safe use of H_2 as an energy carrier.

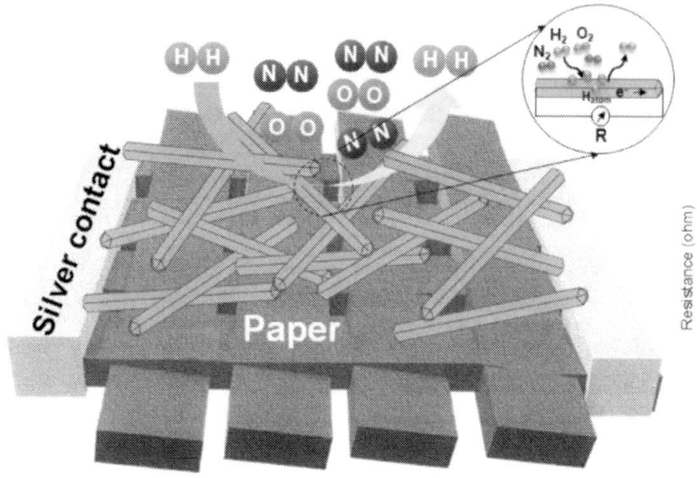
Silver contact
Paper
H₂ O₂
N₂
R

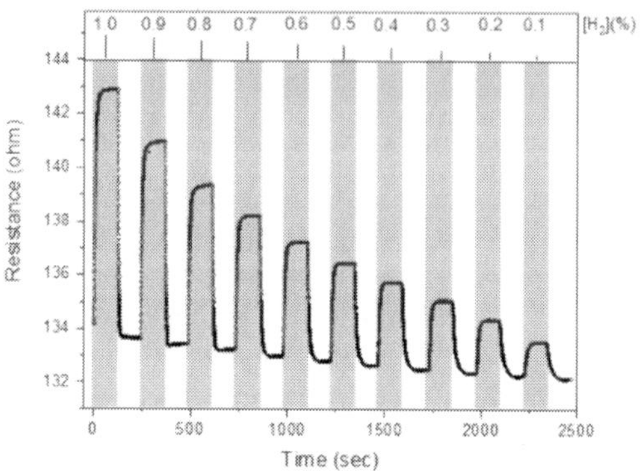
1.0 0.9 0.8 0.7 0.6 0.5 0.4 0.3 0.2 0.1
[H₂](%)
Resistance (ohm)
Time (sec)

626851 Golem: A Probabilistic Approach to Robust Experiment and Process Optimization Based on Regression Trees

Abstract:

Numerous challenges in science and engineering can be framed as optimization tasks, including the maximization of reaction yields, the optimization of molecular and materials properties, and the fine-tuning of automated hardware protocols. Data-driven optimization strategies are often adopted to solve these tasks efficiently. Increasingly, these strategies are coupled with automated hardware to enable autonomous experimental platforms. The majority of the sequential model-based strategies used, however, do not consider robustness against the variability of experiment and process conditions. In fact, it is generally assumed that these parameters are exact and reproducible. Yet some experiments may have considerable noise associated with some of their conditions, and process parameters optimized under precise control may be applied in the future under variable operating conditions. In either scenario, the optimal solutions found might not be robust against input variability, affecting the reproducibility of results and returning suboptimal performance in practice. In this talk, I will introduce *Golem*, an algorithm that is agnostic to the choice of experiment planning strategy and that enables robust experiment and process optimization. *Golem* identifies optimal solutions that are robust to input uncertainty, thus ensuring the reproducible performance of optimized experimental protocols and processes. It can be used to analyze the robustness of past experiments, or to guide experiment planning algorithms toward robust solutions on the fly. We assess the performance and domain of applicability of *Golem* through extensive benchmark studies and demonstrate its practical relevance by optimizing an analytical chemistry protocol under the presence of significant noise in its experimental conditions.

Reference: Aldeghi, M., Häse, F., Hickman, R. J., Tamblyn, I. & Aspuru-Guzik, A. Golem: An algorithm for robust experiment and process optimization. *arXiv:2103.03716* (2021).

628299 Learning Based Scheduling of Industrial Hybrid Renewable Energy Systems

Abstract:

In order to reduce the global CO_2 emissions for low-carbon development while improving the efficiencies of operation, energy systems have started undergoing major transitions to facilitate the broad integration of distributed renewable energy technologies. The inclusion of renewable energy technologies to a typical fossil-fuel based energy source have gradually transformed the concept of centralized generation into distributed energy systems. The choice of optimal power contribution among the available hybrid sources often depends on several factors such as the time of use (TOU) price of the grid electricity, renewable energy generation forecast, electricity demand estimation etc.

Due to the high intermittent nature of renewable energy sources, the system complexity increases, thereby making the optimal decision making more challenging using the existing physics-based model techniques. The inevitable uncertainties related to energy sector such as climate changes, energy market fluctuations, and energy technology developments often demand complex physics-based model and excessive computational time especially for control applications. The physics-based energy scheduling approach generally solves an optimization problem in background to decide the optimal power mix among the available power sources. However, model dependencies, parameter uncertainties and multi-dimensional impact of the embedded uncertainties impose the decision making to be more problematic, especially when such an approach is extended to large-scale and complex energy systems.

The utilization of machine learning techniques in the energy sector has gradually attracted researchers' attention for the effective management of the aforementioned complexities due to their purely data-driven nature. Reinforcement learning (RL), which is a branch of machine learning, can adopt model-free approaches for scheduling under uncertainties by employing agents to participate in the process of interest. The RL agents make control decisions and maximize its total reward by interacting with the simulation environment and evaluating the series of actions taken.

In this study, a data-driven approach that adopts RL techniques to determine the schedule of power mix in a hybrid renewable energy system (HRES) is presented, with a case study on the highly energy intensive chlor-alkali process. This process involves electrolysis of sodium chloride solution to produce chlorine as the primary product together with hydrogen and sodium hydroxide (caustic soda) solution as the secondary products. Almost 50-70% of the overall cost of operating a chlor-alkali process comes from electricity usage, thereby making the process highly sensitive to electricity price trends. To combat the increased grid dependency of such a highly energy intensive process and to achieve profitability, a grid connected hybrid energy system comprising of solar PV panels, wind turbines and fuel cells are considered that supplies power to the chlor-alkali process.

We implement Proximal Policy Optimization (PPO), an Advantage Actor-Critic RL method that enable samples to be trained for multiple epochs of mini-batch updates, to effectively distribute energy among the various available sources without having access to the dynamics of the system. In our exclusive experiments, the RL based scheduling provides efficient operational strategies with saving in cost and show nearly on-par performance with the optimal solution. For stochastic tasks under uncertainties, our RL scheduling approach demonstrates strong superiority over the conventional optimization. This framework can generally be used in any data-driven scheduling problems taking the advantages of its model-free property, and can be generalized to multiple task scenarios via transfer learning.

634107 Keynote Talk: Modelling and Monitoring with Dynamic Auto-Regressive Latent Variable Methods

Abstract:

With the advent of Industry 4.0, industrial data can be collected at very high sampling rates, leading to more complicated relation between process and quality variables[1, 2]. To extract valuable information from these variables and avoid the curse of dimensionality, multivariate latent variable methods (LVM) are widely used, such as partial least squares (PLS)[3], canonical correlation analysis (CCA)[4] and latent variable regression (LVR)[5]. PLS and CCA are employed to construct the relations between process variables **X** and quality variables **Y** by maximizing their covariance and correlation respectively, while LVR aims to maximize the projection of **Y** in the latent space. Several variants of LVM are proposed to achieve better modeling and monitoring performance, including concurrent ones[6-8] and nonlinear versions[9, 10].

The aforementioned models obtain satisfactory modeling and monitoring performance; however, only static variations are considered in these models, and this static assumption constrains their applicability in actual dynamic systems. Several dynamic variants were proposed to extract temporal information from the data. For instance, augmented matrices were formed to include lagged process and quality variables in the work of Ku et al.[11] and Qin and McAvoy[12]. However, it is hard to interpret their latent structures because of the augmentation matrices. To address this issue, dynamic inner LVM was proposed to construct dynamic outer model by maximizing the relations between current samples and a linear combination of past samples, and a dynamic inner model that is consistent with the dynamic outer model[13-16]. Dynamic weighted variants were also proposed in [1, 17] to reduce the redundant information between adjacent historical samples by introducing basis functions. For aforementioned dynamic variants of LVM, only dynamic cross-correlations between the current quality variables and past process and quality variables are considered, leaving large portion of valuable quality information unexploited.

In this work, a series of dynamic auto regressive LVM (DALVM) algorithms are proposed to fully exploit the valuable information of both process and quality variables. DALVM aims to maximize the dynamic relations between current quality samples and past process and quality samples, and to achieve consistency, an auto-regressive exogenous (ARX) model is designed in the inner model. In DALVM, in addition to the auto-regressive relations of quality variables, the cross-correlations between **X** and **Y** are also constructed. Further, aiming for comprehensive modeling and monitoring, subsequent decompositions are proposed to decompose the process and quality variables concurrently into seven dynamic and static subspaces, and the corresponding monitoring scheme is designed for all subspaces. Two industrial dataset, Tennessee Eastman process and the three-phase flow facility, are used to illustrate the advantages of the proposed algorithms.

References

[1] Zhu, Q., 2020. Auto-regressive modeling with dynamic weighted canonical correlation analysis. Journal of Process Control, 95, pp.32-44.

[2] Xu, B. and Zhu, Q., 2020. Online quality-relevant monitoring with dynamic weighted partial least squares. Industrial & Engineering Chemistry Research, 59(48), pp.21124-21132.

[3] Geladi, P. and Kowalski, B.R., 1986. Partial least-squares regression: a tutorial. Analytica chimica acta, 185, pp.1-17.

[4] Hardoon, D.R., Szedmak, S. and Shawe-Taylor, J., 2004. Canonical correlation analysis: An overview with application to learning methods. Neural computation, 16(12), pp.2639-2664.

[5] Zhu, Q., 2020. Latent variable regression for supervised modeling and monitoring. IEEE/CAA Journal of Automatica Sinica, 7(3), pp.800-811.

[6] Qin, S.J. and Zheng, Y., 2013. Quality-relevant and process-relevant fault monitoring with concurrent projection to latent structures. AIChE Journal, 59(2), pp.496-504.

[7] Zhu, Q., Liu, Q. and Qin, S.J., 2017. Concurrent quality and process monitoring with canonical correlation analysis. Journal of Process Control, 60, pp.95-103.

[8] Zhang, H. and Zhu, Q., 2021. Multi-layer fault monitoring with concurrent kernel latent variable regression. Submitted to Industrial & Engineering Chemistry Research.

[9] Melzer, T., Reiter, M. and Bischof, H., 2003. Appearance models based on kernel canonical correlation analysis. Pattern recognition, 36(9), pp.1961-1971.

[10] Zhang, Y., Zhou, H., Qin, S.J. and Chai, T., 2009. Decentralized fault diagnosis of large-scale processes using multiblock kernel partial least squares. IEEE Transactions on Industrial Informatics, 6(1), pp.3-10.

[11] Ku, W., Storer, R.H. and Georgakis, C., 1995. Disturbance detection and isolation by dynamic principal component analysis. Chemometrics and intelligent laboratory systems, 30(1), pp.179-196.

[12] Qin, S.J. and McAvoy, T.J., 1996. Nonlinear FIR modeling via a neural net PLS approach. Computers & chemical engineering, 20(2), pp.147-159.

[13] Dong, Y. and Qin, S.J., 2018. Dynamic latent variable analytics for process operations and control. Computers & Chemical Engineering, 114, pp.69-80.

[14] Dong, Y., Liu, Y. and Qin, S.J., 2019. Efficient dynamic latent variable analysis for high-dimensional time series data. IEEE Transactions on Industrial Informatics, 16(6), pp.4068-4076.

[15] Zhu, Q., Qin, S.J. and Dong, Y., 2020. Dynamic latent variable regression for inferential sensor modeling and monitoring. Computers & Chemical Engineering, 137, p.106809.

[16] Qin, S.J., Dong, Y., Zhu, Q., Wang, J. and Liu, Q., 2020. Bridging systems theory and data science: A unifying review of dynamic latent variable analytics and process monitoring. Annual Reviews in Control.

[17] Xu, B. and Zhu, Q., 2020. Online Quality-Relevant Monitoring with Dynamic Weighted Partial Least Squares. Industrial & Engineering Chemistry Research, 59(48), pp.21124-21132.

633493 Investigation of Electrochemical Redox Tags for Elastin-like Polymer Sensing

Abstract:

Elastin-Like Polymers (ELP) have been investigated as surface constructs to promote surface interactions or as a stimuli-responsive surface.[1,2] We have previously demonstrated using electrochemical impedance spectroscopy to detect an ELP collapse response.[1] However, EIS measurements are exceptionally sensitive to minute changes in solution or surface conditions, which makes consistent data acquisition difficult.[3] In order to create a more robust signal to the same collapse response, we propose using an electrochemical tag coupled with voltametric techniques to recognize the surface response. Differences in electrochemical characteristics of these compounds in gold electrodes versus glassy carbon electrodes will be the focus of this poster.

1. Su, Z., Kim, C. O. and Renner, J. N. Quantification of the effects of hydrophobicity and mass loading on the effective coverage of surface-immobilized elastin-like peptides. *Biochem. Eng. J.* **168**, 107933 (2021). 10.1016/j.bej.2021.107933

2. Morales, M. A., Paiva, W. A., Marvin, L., Balog, E. R. M. and Halpern, J. M. Electrochemical characterization of the stimuli-response of surface-immobilized elastin-like polymers. *Soft Matter* **15**, 9640–9646 (2019). 10.1039/c9sm01681c

3. Bogomolova, A., Komarova, E., Reber, K., Gerasimov, T., Yavuz, O., Bhatt, S. and Aldissi, M. Challenges of Electrochemical Impedance Spectroscopy in Protein Biosensing. *Anal. Chem.* **81**, 3944–3949 (2009). 10.1021/ac9002358 https://pubs.acs.org/doi/10.1021/ac9002358

629889 Mammalian Cell-Compatible "Light-up" RNA Aptamer Biosensors for Custom Small Molecule Targets

Abstract:

In vivo small molecule imaging is an exciting area in biosensor research. Molecular imaging has applications in investigating and visualizing biological processes, screening and diagnosis, personalized medicine, and drug development. Currently a variety of methods are used both clinically and pre-clinically including CT, PET, SPECT, MRI, and ultrasound. Additionally, bioluminescence, fluorescence, photoacoustic, and Raman based techniques are also used in research efforts. Still, improvements to spatial, temporal, and molecular specificity and sensitivity are desired.

The use of aptamers poses a potential solution by using single-stranded DNA or RNA oligonucleotides capable of binding to specific molecules. Efforts to use fluorescent aptamers for direct small molecule sensing were pioneered with the development of Spinach, an RNA aptamer mimic of GFP (green fluorescent protein). Spinach was subsequently converted to a fluorescent small molecule sensor by inserting naturally-occurring riboswitch motifs into the parent fluorescent RNA structure.

Spinach was originally generated using standard Systematic Evolution of Ligands by Exponential Enrichment (SELEX) approaches to isolate aptamers that could bind the DFHBI chromophore. A revisit to the SELEX process was used to select newer aptamers via expression of RNAs within *E. Coli* cells and fluorescence-activated cell sorting (FACS). This method was used to create Broccoli, a markedly improved aptamer with nearly twice the brightness of Spinach, enhanced thermostability, and ion biocompatibility.

Overall, despite these advances in the use of fluorescent aptamers for molecular detection in mammalian cells, a persistent limitation still exists with the small molecule recognition motifs. Specifically, recognition motifs to this point have all been derived from naturally-occurring riboswitches, which limits the applicability of these techniques to small molecules with cognate riboswitch pairs. This limitation also completely eliminates the use of aptamer biosensors for non-native small molecules like synthetic drugs. SELEX techniques have been developed to generate aptamers with riboswitch-like activity. For example, a technique called capture SELEX selects for conformational changing aptamers. However, such SELEX techniques for generating structure-switching motifs have not been combined with aptamer biosensor workflows, which would enable new capabilities in this space.

Here, we sought to develop an integrated SELEX approach to address this issue. We focused on l-3,4-dihydroxyphenylalanine (L-DOPA), a normally occurring amino acid precursor to the neurotransmitter dopamine, for which an existing riboswitch does not exist or has not been identified. We used capture SELEX to generate RNA candidates that could prospectively bind L-DOPA. We inserted a subset of candidates into the Broccoli construct and tested fluorescence in transfected HEK cells in the presence and absence of L-DOPA. These tests led to the identification of aptamer 3F, which exhibited modest light-up activity in cells. Live cell imaging was performed on Caco2 cells transfected with 3F in the same expression system. Caco2 cells differ from HEK cells in that they have the enzyme dopa decarboxylase responsible for the metabolism of L-DOPA. This enzyme can be inhibited by carbidopa. 3F transfected Caco2 cells given L-DOPA and DFHBI steadily lost fluorescence over time while those treated with carbidopa maintained fluorescence 45 minutes later.

In general, L-DOPA metabolism causes fluorescence to decrease while inhibition of this degradation pathway maintains higher L-DOPA concentration and fluorescence. This data suggests that 3F fluorescence is dependent on L-DOPA concentration in mammalian cells. To overcome inefficient transfection efficiencies, we created a stable 3F-expression Caco-2 cell line and carried out detailed characterizations of L-DOPA biosensing dynamics. Overall, our results establish a strategy for creating mammalian cell-compatible RNA biosensors against custom small molecule targets.

630933 Microfabriction of GEM Sensors for Quantitative Glucose Mapping and Multiplexing with MRI

Abstract:

Magnetic resonance imaging (MRI) has widespread clinical application due to its operation in the radio frequency range where distortion from biological tissue is negligible, but generally lacks quantitation and multiplexing strategies. In this work, we present recent progress in microfabricating glucose-responsive Geometrically Encoded Magnetic (GEM) sensors, a new sensor platform with inherent advantages to address the quantitation and multiplexing issues with MRI. GEM sensors are composed of size-changing 'smart' hydrogels that are infused with magnetic nanoparticles. The magnetic nanoparticles create a local magnetic field which in turn produces a resonance frequency offset that distinguishes local hydrogel molecules from surrounding ones on a nuclear magnetic resonance (NMR) spectrum. The smart hydrogel reversibly swells and contracts in response to analyte concentration, actuating a change in the sensor shape, a change in the strength of the locally created magnetic field, and a shift in the offset frequency. Analyte concentration can therefore be read from the relative position of the frequency shift from a starting point, giving an inherently ratiometric signal readout that is not subject to artifacts from sensor concentration. While the GEM sensor concept has been demonstrated with a pH-responsive hydrogel, current efforts are aimed at expanding the range of accessible analytes by determining the proper conditions for GEM sensor microfabrication with a glucose-responsive hydrogel formulation through a cost-effective soft lithography approach. In the future, by constructing GEM sensor shapes with different aspect ratios that correspond to various hydrogel chemistries, a suite of sensors can be built where each analyte concentration corresponds to its relative shift in distinct regions of the NMR spectra, analogous to multiplexing with optical tags.

624569 A Wearable Patch for Sweat Lactate Analysis

Abstract:

Sweat lactate is an essential biomarker for monitoring anaerobic metabolism, oxidative stress levels, muscle health, and tissue hypoxia in individuals. Its concentration depends on the duration and the intensity of the physiological state (rest or exercise) that the body experiences. Most of the current sweat sensing devices measure sweat lactate for a short duration (~ minutes) and are incapable of delivering comprehensive data on lactate metabolism in the body. We address these issues with our non-invasive, zero-electric powered wearable sweat sampling patch[1], that can measure sweat lactate levels under rest, during medium intensity exercise, and post-exercising conditions upon directly interfacing with the skin. Our patch simultaneously uses osmosis, capillary wicking, and evaporation to deliver long-term sweat collection and lactate monitoring. Osmotic action is achieved through a hydrogel with a higher solute content than sweat, while wicking and evaporation occur in paper microfluidic channel. We use a glucose hydrogel, PBS hydrogel, and a silicone disk as the major pumping sources in the patch and test them under different physiological conditions. On body trials show that all three pumping materials can register lactate on paper for 2 hours. The glucose hydrogel withdraws sweat lactate majorly via osmosis during rest and post-exercise trials. Sampling of lactate via osmosis reduces upon exercising due to inflow of active sweat in the paper channel. The PBS hydrogel and silicone disk patches show traces of lactate on paper as an outcome of the natural sweat release by the body. We also measured blood lactate levels and found no correlation between blood and sweat lactate levels under all physiological conditions. Hence, sweat appears to be a more informative medium than blood for lactate. Overall, such simple, non-invasive patches for measuring sweat lactate levels can generate a wide variety of information about human health and contribute towards the advancement of next generation wearable health monitoring devices.

References:

1. Saha, T., Fang, J., Mukherjee, S., Dickey, M.D., and Velev, O.D. (2021) Wearable Osmotic-Capillary Patch for Prolonged Sweat Harvesting and Sensing. *ACS Appl. Mater. Interfaces*, **13** (7), 8071–8081.

625215 Evaluating the Fitness of Combinations of Adsorbents for Quantitative Gas Sensor Arrays

Abstract:

Robust, high-performance gas sensing technology has applications in industrial process monitoring and control, air quality monitoring, food quality assessment, disease diagnostics, and security threat detection. Metal-organic frameworks (MOFs)-- a class of nanoporous materials with large, functionalizable internal surface areas-- are promising recognition elements for gas sensors, owing to their selective gas adsorption properties. Mimicking mammalian olfaction, the response pattern of an *array* of diverse MOFs could enable inference of the composition of complex, multicomponent gas mixtures. As modular and tunable materials, MOFs offer a vast materials space to search for diverse combinations to compose a robust gas sensor array.

We propose a mathematical method to evaluate the fitness of combinations of MOFs, based on their adsorption properties, for quantitative gas sensor arrays. This method enables the virtual screening of MOF-based gas sensor arrays, moving from trial-and-error sensor array development to *design*. First, we frame gas sensing as an inverse problem. While the (routine) forward problem is to predict the mass of gas adsorbed in each MOF when immersed in a gas mixture of a given composition, the inverse problem is to predict the gas composition from the observed masses of adsorbed gas in the MOFs of the array. The fitness of a given combination of MOFs for gas sensing is then determined by the conditioning of its inverse problem: the prediction of the gas composition provided by a fit (unfit) combination of MOFs is insensitive (sensitive) to inevitable errors in the measurements of the mass of gas adsorbed in each MOF.

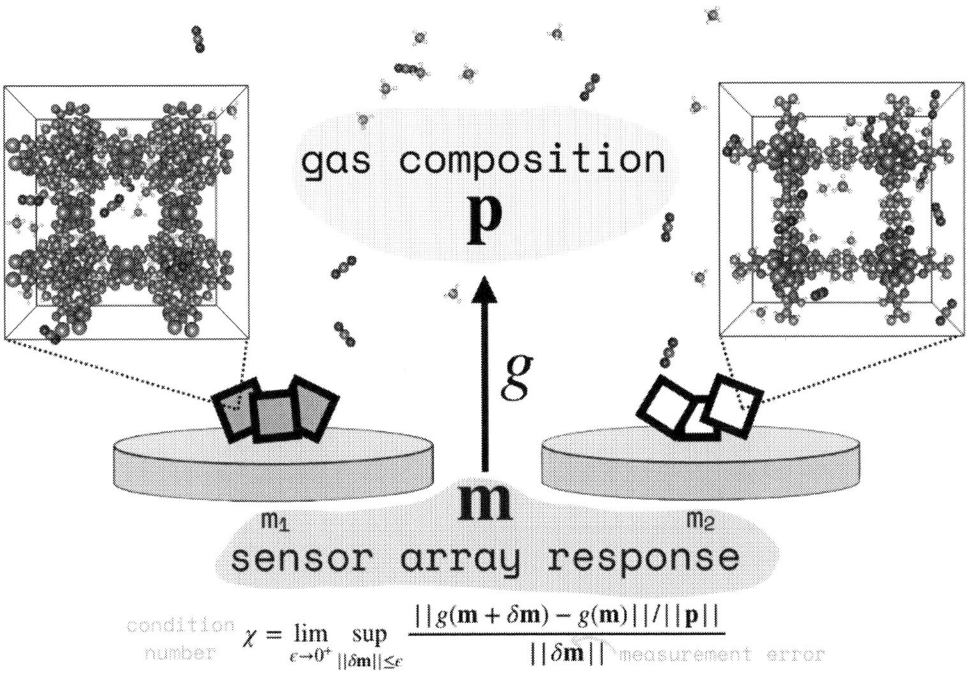

634114 Keynote Talk: Online Measurements for the Petrochemical Industry: Industry 4.0 Trends and Unmet Needs

Abstract:

The primary objectives of the petrochemical manufacturing industry are to operate plants safely, profitably, and reliably. To achieve these objectives, several constraints need to be managed – such as raw material yield, manufacturing costs, and utility requirements. Online measurements via instruments that measure the physical properties and analyzers that measure the chemical properties of the process are critical to the optimized management of these constraints.

Globally, Dow owns and supports > 1 million instruments and > 20,000 online analyzers. Data from these measurements are indispensable to the operation and control of our plants. In this talk we will discuss some of our key requirements for both classes of online measurements, share examples of meaningful advances afforded by Industry 4.0, and articulate key challenges that remain – a call for continued innovation.

628357 Woodchip Moisture Content Estimation Using Short-Range Iot Wi-Fi for the Pulp & Paper Industry

Abstract:

The pulp and paper industry is the third-largest consumer of energy in the US industrial sector thereby leading to tremendous opportunities to improve its energy efficiency and productivity. The pulping process, which converts woodchips into pulp by displacing lignin from cellulose fibers, is one of the most important operations in a pulp and paper mill. For the pulping process, it is important to have real-time knowledge about the moisture content (MC) of the woodchips so that the process can be optimized and/or controlled correspondingly to achieve satisfactory product quality while minimizing the consumption of energy and chemicals. There have been attempts to develop real-time woodchip MC sensors, such as capacitance, resistance, microwave and NIR-based sensors. But these techniques have not been adopted by the industry due to unsatisfactory performance and/or high maintenance requirements that can hardly be met in a manufacturing environment. Also, some of the accurate methods are often lab-based that cannot be implemented online because of time constraint [1]. Due to these limitations, at present, the moisture level is measured only a few times throughout the year in the pulp and paper industry.

To address these limitations, we propose a non-destructive, economic, and robust approach using industrial Internet-of-Things (IIoT) based Wi-Fi and employ channel state information (CSI) to estimate MC in woodchips. We extract CSI by modifying the open-source device drivers for Intel Wi-Fi link 5300 Network Interface Card based on CSITool [2]. CSI contains information about the channel in the form of individual data sub-carriers capturing indoor channel features such as the effect of scattering, fading, and power decay with distance. We hypothesize that when a Wi-Fi signal passes through wood chips, the water or MC in woodchips would cause a corresponding change in the Wi-Fi signal which is then captured in CSI. The goal is to develop a data-driven model to extract the relationship between the change in CSI and the woodchip MC. CSI has been successfully used for moisture classification in wheat [3]. However, compared to wheat moisture classification, woodchip moisture estimation is much more challenging due to the significant heterogeneity in the shape, structure, and arrangement of the woodchips. For example, the woodchip arrangement in the container is expected to have a significant impact on the CSI data. In addition, CSI data have only been used in classifications, which is much easier than regression modeling.

In this work, we propose a regression-based approach to estimate the moisture content in wood chips. To the best of our knowledge, this is the first work to address the continuous estimation of MC using CSI data. Our goal is to develop a robust model that is insensitive to the heterogeneity of the woodchips, particularly the confounding factors such as woodchip shape and the effect of shuffling, to achieve a consistent and accurate MC estimation. While the IIOT devices used are small, low-cost, and rugged to stand for harsh environment, they do have their limitations such as the raw CSI data are very noisy, which can lead to low performing models when directly fed to machine learning algorithms. To address this challenge, our previously developed statistics pattern analysis (SPA) framework [4][5] is used to extract robust and predictive features from raw CSI data. The SPA-based features are then used to develop robust machine learning models for continuous MC estimation in woodchips. In other words, SPA turns messy big data into so-called smart data, which can then be used for accurate sensing. The machine learning techniques studied in this work include artificial neural networks (ANN), support vector machines (SVM), and Gaussian process regression (GPR). We conducted systematic study of these nonlinear regression approaches, and demonstrated that, when combined with SPA, provide a potential solution for real-time woodchip MC estimation for the pulp and paper industry. Specifically, the experimental validation results show highly accurate prediction of more than 20 MC levels in a wide range. In addition to the excellent performance, the proposed approach is also economic and non-destructive, making it a potential technology for industrial implementation in the pulp and paper industry.

[1] R. Govett, A PRACTICAL GUIDE FOR THE DETERMINATION OF MOISTURE CONTENT OF WOODY BIOMASS A Practical Handbook of Basic Information, Definitions, Calculations, Practices and Procedures for Purchasers and Suppliers of Woody Biomass.

[2] D. Halperin, W. Hu, A. Sheth, and D. Wetherall, Tool Release: Gathering 802.11n Traces with Channel State Information, *ACM SIGCOMM Comput. Commun. Rev.*, 2011.

[3] W. Yang, X. Wang, S. Cao, H. Wang, and S. Mao, Multi-class wheat moisture detection with 5GHz Wi-Fi: A deep LSTM approach, in *Proceedings - International Conference on Computer Communications and Networks, ICCCN*, 2018.

[4] Q. P. He and J. Wang, Statistics pattern analysis: A new process monitoring framework and its application to semiconductor batch processes, *AIChE J.*, 2011.

[5] J. Wang and Q. P. He, Multivariate Statistical Process Monitoring Based on Statistics Pattern Analysis, *Ind. Eng. Chem. Res.*, 2010.

633025 Gradient-Weighted Class Activation Mapping (Grad-CAM) Based Explanations for Process Monitoring Results from Deep Neural Networks

Abstract:

In the chemical process industries, high product quality with maximum efficiency and process safety are the critical issues that require effective process monitoring. A fault in a process results in the deviation of process state variables from its normal operating condition. The effect of the fault propagates along the process and disturbs stability of the system. So, it is essential to detect them quickly on their occurrence. To tackle this problem, we motivate to develop the methodology which fulfils the aspect of high accuracy, novel explainability and to make effective use of the multivariate time series process data.

Recent developments in sensor techniques and storage technologies have provided a significant technology push for developing next-generation data driven process monitoring methods [1]. Hence deep learning methods have attracted much attention for fault detection. Deep learning methods like Deep Belief Network (DBN), Convolutional Neural Networks (CNN) and, Hierarchical Deep Neural Networks (HDNN) have been used for fault detection [2] but they lack in providing explanations for their decision. Hence, the eXplainable Artificial Intelligence (XAI) method Integrated Gradients was used by [3] to provide explanations for the results generated from deep learning models. But the Integrated Gradient suffer from one shortcoming that it analyses the samples one by one without accounting for the fact that the data is a time series. In the paper, we develop the XAI method which accounts for the time series nature of the process monitoring problem.

Recently, a new XAI method called Grad-CAM has been developed targeted at providing explanations for time series data in terms of activation maps for CNN [4]. Grad-CAM helps to explore the spatial information that is preserved inside convolution layers. The extracted information from CNN layers helps to understand the important feature of an input data to make classification decision. Grad-CAM uses k feature maps of height h and width u of the last convolution layer, as it contains high level semantics and detailed spatial information. For each feature map A^k, we obtain an importance weight w_k^c associated with specific class c. The w_k^c is obtained by computing the gradient of final class output y^c with respect to the feature map A^k i.e., $\partial y^c / \partial A^k$. It is further globally averaged over width u and height h. This step is followed by computing a weighted sum of feature maps A^k of class c using w_k^c. The final heatmap is obtained by applying the *ReLU* operation on the weighted sum to emphasize only positive values. This heatmap highlights the features which are important in making a final decision of the network for class c. In this paper, we use the Grad-CAM method for the explanation of process fault diagnosis.

In this paper, we exploit the real time explainable CNN methodology that classifies the normal and faulty state of the process. The proposed methodology consists of a 2D Convolution layer that is trained to solve a fault detection problem. Then, Grad-CAM is used on a trained model to understand which part of input data contributes more to detect a fault [5]. The contribution of different features visualized using the generated heatmap for each sample helps to identify the key factors in fault occurrence. The performance of the proposed methodology demonstrated on simulated time series classification datasets which contain different input variables and categorical outputs. The CNN was trained and heatmaps were generated using Grad-CAM that provide an accurate explanation about the predicted outcome. In this paper, we will describe the proposed methodology as well as results obtained from the simulated dataset which effectively makes use of the time series dimension to provide intuitive explanations for outputs. The results from fault diagnosis on the Tennessee Eastman challenge problem will also be demonstrated [6].

References

[1] P. Park, P. D. Marco, H. Shin, and J. Bang, "Fault Detection and Diagnosis Using Combined Autoencoder and Long Short-Term Memory Network," *Sensors*, vol. 19, no. 21, p. 4612, Oct. 2019, doi: 10.3390/s19214612.

[2] H. Wu and J. Zhao, "Deep convolutional neural network model based chemical process fault diagnosis," *Comput. Chem. Eng.*, vol. 115, pp. 185–197, Jul. 2018, doi: 10.1016/j.compchemeng.2018.04.009.

[3] V. Pakkiriswamy and R. Srinivasan, "An Explainable Artificial Intelligence Based Approach for Interpretation of Fault Detection Results from Deep Neural Networks," presented at the 2020 Virtual AIChE Annual Meeting, 2020, [Online]. Available: https://www.aiche.org/academy/conferences/aiche-annual-meeting/2020/proceeding/paper/375g-explainable-artificial-intelligence-based-approach-interpretation-fault-detection-results-deep.

[4] R. R. Selvaraju, M. Cogswell, A. Das, R. Vedantam, D. Parikh, and D. Batra, "Grad-CAM: Visual Explanations from Deep Networks via Gradient-based Localization," *ArXiv161002391 Cs*, Dec. 2019, doi: 10.1007/s11263-019-01228-7.

[5] R. Assaf and A. Schumann, "Explainable Deep Neural Networks for Multivariate Time Series Predictions," in *Proceedings of the Twenty-Eighth International Joint Conference on Artificial Intelligence*, Macao, China, Aug. 2019, pp. 6488–6490, doi: 10.24963/ijcai.2019/932.

[6] J. J. Downs and E. F. Vogel, "A plant-wide industrial process control problem," *Comput. Chem. Eng.*, vol. 17, no. 3, pp. 245–255, Mar. 1993, doi: 10.1016/0098-1354(93)80018-I.

630074 A Deep Learning Vision System for Classification of Manufacturing Defects

Abstract:

Manufacturing lines for the assembly of pharmaceutical autoinjectors are highly automated and rely on visual inspection stations to efficiently and safely process a large volume of syringes. As pre-filled syringes enter the inspection line, there are a variety of subtle defects in the flange (e.g., chips, cracks) as well as other processing flaws (e.g., missing components) that can make the assembly unfit for use or otherwise problematic for continued processing. Currently, there is a highly-engineered inspection algorithm that is manually tuned for identifying defects from camera images. The existing approach is able to robustly detect important defects, however, a relatively large percent of rejected products are in fact false positives (e.g., cosmetic defects, lighting irregularities). In this work, we have developed a convolutional neural network approach for classifying defects in syringes with comparable sensitivity and higher selectivity than the previous algorithm. We found that preprocessing and noise reduction were critical for training meaningful models from available images. Additionally, we highlight a series of infrastructure challenges that were overcome during image acquisition and algorithm deployment to transfer our work to and from a good manufacturing process (GMP) inspection line. We also illustrate the importance of algorithm-architecture co-design when developing machine learning models in order for the solution to be practically deployed on the available edge computing device (i.e., the vision processor). This work highlights the use of machine learning in conjunction with traditional computer vision to classify defects beyond what can be identified by expert intuition.

631951 Rapid Semiconductor Sensor System for Cerebral Spinal Fluid Detection Using Disposable Biosensor Strips

Abstract:

Cerebral spinal fluid (CSF) is an extracellular fluid surrounding the brain and spinal cord for nutrition transport and physical buffer. The leakage of cerebral spinal fluid is of crucial clinical concern with respect to the possibility of potentially fatal infections. Current test method for beta-2-transferrin (B2T), a biomarker for endogenous protein only present in CSF, requires hours of turnaround time in centralized laboratory and can potentially delay the diagnosis and timely treatment. Semiconductor based sensor offers the unique advantage of providing fast detection as well as possibilities of modular packaging for point-of-care use. Si metal oxide semiconductor field effect transistor (MOSFET) was implemented for signal amplification purpose, instead of expensive high electron mobility transistor (HEMT). The gate electrode of the transistor is externally connected to a biofunctionalized sensing strip with beta-2-transferrin (B2T) monoclonal antibodies (mAb). Compare to previous generation of sensor that requires complex fabrication steps on glass slides, this work presents the use of factory-premade plastic sensor for cost reduction, improved production throughput, while achieving high sensitivity. With this sensor system, detection of cerebrospinal fluid can be achieved range from 0.1ng/mL to 10mg/mL with one time use sensor strips biofunctionalized with B2T mAb, as well as detection time of less than 1 second. Immediate feedback of presence of CSF leakage via medical sensor lessened the turnaround time compared to traditional laboratory test, and this life-saving technology can be potentially implemented in detection of board range of biomarkers of interest.

623689 Realistic Thermal Sensor Modeling Using Finite Element Method (FEM) of Direct Metal Laser Solidification (DMLS)

Abstract:

In the additive manufacturing (AM) industry, the direct metal laser solidification (DMLS) process has received significant research interest because of its ultra-high precision and geometry variability [1]. However, because of the thermal and mechanical complexity, process failures are often encountered in DMLS, which maybe lead to part defects or even detrimental damage to the printing platform. These heating abnormalities may lead to thermal and mechanical stresses on the build part and eventually may cause physical problems like keyholing or lack of fusion [2]. As a result, crucial problems like recoater jam may be induced by the undesired part extrusion above the powder surface [3]. To visualize what is happening during the printing process and, in-situ sensors are developed to investigate and record the printing process information [4].

Thus, to facilitate engineers to understand the abnormality and the underlying physics of the complicated monitoring output, a large-scale finite element method (FEM) model is developed to investigate the heat transfer behavior of the DMLS process and extract experimentally relevant thermal features. Specifically, a microscopic and a meso-level sub-model are initially developed to describe powder properties and the laser behaviors, respectively, and their outputs are directly incorporated in the experimental-scale FEM model. The FEM model-generated data are then processed to reproduce the long-wave infrared camera (LWIR) sensor outputs to relate the physical defects and the output thermal images.

[1] Liu, R., Wang, Z., Sparks, T., Liou, F., Newkirk, J., 2017. Aerospace applications of laser additive manufacturing, in: Laser Additive Manufacturing. Elsevier, pp. 351–371.

[2] Grünberger, T., Domröse, R., 2014. Optical in-process monitoring of direct metal laser sintering (DMLS): A revolutionary technology meets automated quality inspection. Laser Technik Journal11, 40–42.

[3] Gong, H., Rafi, K., Gu, H., Starr, T., Stucker, B., 2014. Analysis of defect generation in Ti–6Al–4Vparts made using powder bed fusion additive manufacturing processes. Additive Manufacturing1, 87–98.

[4] Scott, C.,2017.EOS introduces EOSTATE Exposure OT, First Commercial Optical Tomography System for Additive Manufacturing. https://3dprint.com/178624/eos-eostate-exposure-ot/(Accessed on 2020-07-19).

623595 Detection of Methane Gas Using Surface Acoustic Wave Sensors Coated with a Composite Zeolite Layer

Abstract:

Natural gas is widely utilized as a source to generate energy and as a feedstock to manufacture chemicals and materials. The U.S. natural gas pipeline network consists of approximately 305,000 miles of interstate and intrastate transmission pipelines, which is about 12 times the Earth's circumference. The estimated methane emissions from the U.S. natural gas pipelines are 0.69 Tg/year, equivalent to 2.3% of gross U.S. gas production and a loss of ~$2 billion a year. As a powerful greenhouse gas, methane leaks present serious concerns to climate change while also increasing global levels of surface ozone pollution. Thus, it is imperative to first detect leaks from natural gas infrastructure using sensitive and reliable methane sensors, and afterwards to repair leaks and upgrade aging pipelines that have become prone to leaks. In this presentation, surface acoustic wave (SAW) sensors coated with a composite zeolite layer were developed to detect methane gas at various concentrations. The effect of optimizing variables such as particle size, zeolite percentage, binder contents, etc. on the methane sensing response will be discussed, and any cross talk with humidity and other gases will also be presented. The SAW sensors developed here enable the detection of methane leaks along natural gas pipelines in a sensitive and selective manner, ultimately resulting in improvements in natural gas utilization and the reduction of negative environmental impacts.

623033 Novel Technology for Studying Mixing and Reaction State Using Electrical Resistance Tomography

Abstract:

The accurate determination of endpoints of reactions is crucial in the process industry, directly influencing the final product quality or downstream separation processes. Many existing technologies focus on offline sample testing. Due to this kind of testing being a single time point and a single location, in most cases, confidence in the outcome needs to be gained through repeated testing using multiple locations. Electrical resistance tomography (ERT) has been applied in a wide range of scientific and engineering fields, such as geophysical study, clinical diagnosis, chemical or mineral process measurement, and process optimization. ERT is usually used to collect data on the processes in vessels and pipes. In this research, ERT uses an array of 16 electrodes located equidistant in a layer around the inside of the vessel. As fluids flow past the measurement plane, conductivity is measured in the 2D plane. Its ability to do online real-time measurements in a plane and without the need to draw samples from the reaction vessel made the data collection much easier compared to existing technologies. The information on the state of the reaction and the fluid mixing can be extracted from the ERT measurements.

Our ERT system can collect more than 1000 datasets per second with 104 data points per dataset. These real-time datasets contain rich information that can be extracted by data processing. The test case of titration is currently done in a small vessel that acts as a batch reactor, which provides the best similarity to the future application. 0.1 molar hydrochloric acid is titrated with 1.0 molar sodium hydroxide to produce sodium chloride. This reaction has an extensive conductivity range and provides the best case for identifying the reaction's endpoint. Figure 1 shows an online real-time ERT measurement to detect the titration reaction's endpoint at 88.0 seconds. When compared to other methods to find the endpoint of the reaction, it can be seen that the start of the upward trend after the dip matches up closely to those measured endpoints. Additionally, more accurate time points can be obtained by the ERT system comparing to other methods. Another significant benefit of the 2D planar measuring is the ability to monitor the fluid mixing. Figure 2 reveals a location with slower mixing relative to the rest of the vessel. Knowing where these spots exist allows for monitoring the products for any potential defects due to this imperfect mixing. The future work is testing larger-scale vessels and a variety of chemical reactions. This novel technology can be potentially used in large reactors where visibility, testing access, or conditions restrict access to the vessel, such as higher pressure and temperatures.

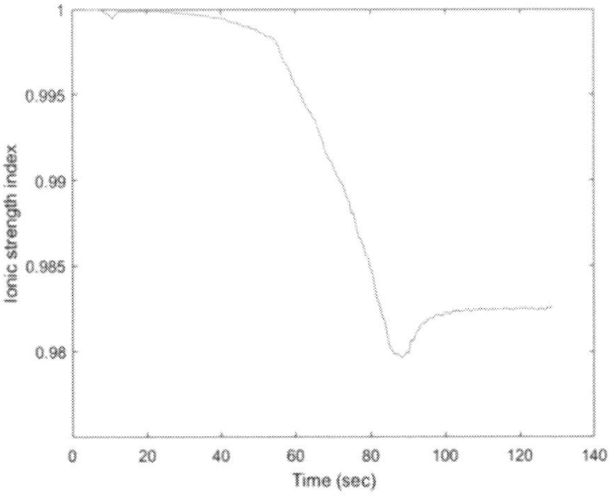

Figure 1: 0.1 Molar HCl titrated with 1.0 Molar NaOH

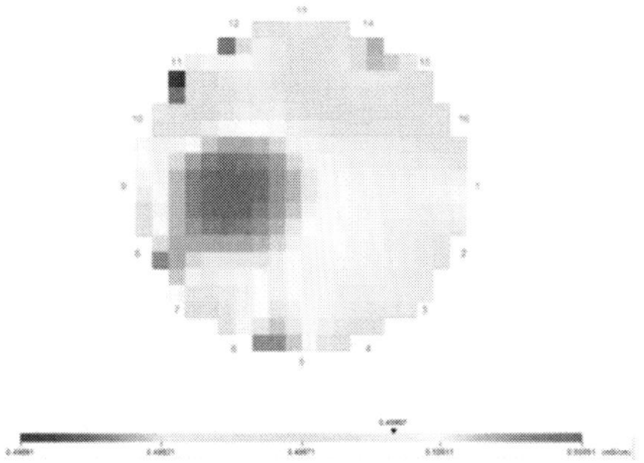

Figure 2: Spot of less mixing.

632478 Flexible Hydrogel-Based Biosensor Fabricated Via Direct Laser Writing for Detection of Cortisol

Abstract:

Wearable sensors have attracted much attention thanks to their promising applications in personalized medicine through remote monitoring of physiological parameters. Cortisol is a biomarker for numerous diseases and is important for blood pressure regulation, glucose levels and metabolism. This study is to create a flexible, wearable biosensor to detect cortisol in body fluid by skin contact with the sensor. A laser-assisted processing technique was used to produce laser-induced graphene (LIG) by direct laser writing with a carbon dioxide laser on carbon materials. This LIG substrate was transferred to an elastomeric substrate (PDMS). For non-invasive extraction of biomarkers, a hydrogel layer was incorporated onto the electrode surface. A cortisol antibody was immobilized on the working electrode via surface chemistry. Both cyclic voltammetry and electrochemical impedance spectroscopy were used to characterize the modified electrode and perform detection of cortisol. Specifically, the characterizations were carried out in terms of detection limit, selectivity, sensitivity, response time, and linear range. Using 3D printed wristbands and wireless communication devices, it may be possible to monitor personnel in real-time during any sort of situations including long-term exposure to space environments and solid contact during physical activity.

628413 *A Visible Colorimetric Sensor Based on Hydrophobic Syndiotactic Polypropylene Fiber Mats for the Determination of Trihalomethanes in Treated Drinking Water*

Abstract:

Water is considered by many to be the most imperative resource on the planet. Despite its essentialness for life, water frequently undergoes pollution from many human processes, most notably industrial runoff. As a result, hazardous chemicals can be found in water sources, and in some cases, even drinking water. One specific class of carcinogenic chemicals of interest are trihalomethanes (THMs), which occur as a byproduct of water disinfection via chlorination, where natural organic matter reacts with halogens to form these toxic chemicals. The US Environmental Protection Agency (USEPA) limits the total concentration of total-THMs in drinking water to 80 ppb. Here, a colorimetric polymer fiber-based sensor was used for the detection of THMs down to 8 ppb. The sensor utilizes the Fujiwara reaction where THMs react with pyridine, which under basic conditions forms red chromophores whose intensity can be measured using image analysis software. The sensor operates by using a two-fold preconcentration technique consisting of simple thermodynamic separation and a hydrophobic syndiotactic polypropylene (sPP) mat to prevent the diffusion of evaporated water through the fibers, as water is detrimental to the Fujiwara reaction. This technique is inexpensive and more rapid than typical THM detection methods currently used to regulate the THMs in water supplies and can be performed on site. As this method is improved, it will be able to be used to ensure that water is potable and safe for drinking.

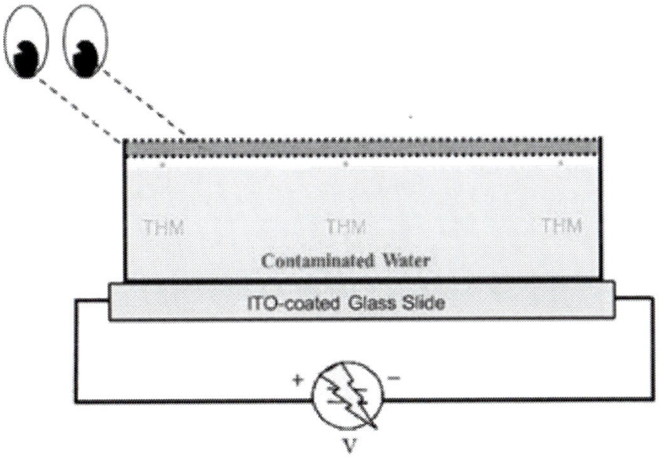

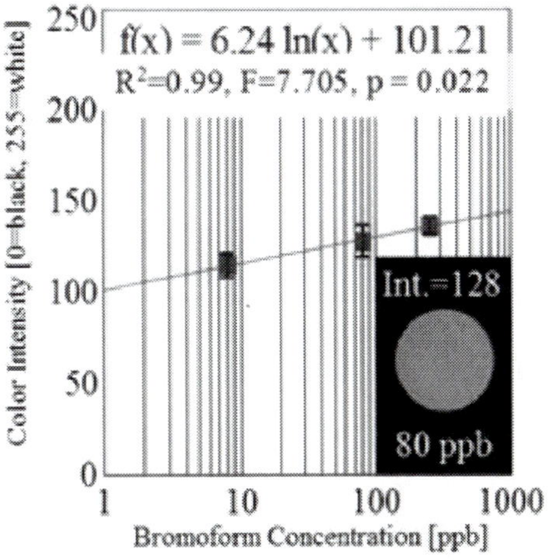

624330 Microneedle-Assisted Detection of Plant Pathogens on a Smartphone

Abstract:

Early detection of plant diseases and timely determination of treatment strategies require a paradigm shift of molecular diagnosis from the laboratory to directly in the field (*Nanoscale Adv*. 2020, 2, 3083-3094). The current molecular diagnostic methods are inefficient to monitor the disease outbreak and spread over time, as the diagnostic tests are either performed in a laboratory setting (e.g., nucleic acid amplification) or lack sensitivity and specificity (e.g., lateral flow test strips). This work describes a rapid solution for in-field diagnosis of plant diseases by combining MN extraction technology with a smartphone-enabled nucleic acid diagnostic platform for quick amplification, quantification, and result reporting all in one device (*Biosensors and Bioelectronics*, under revision). The MN patch extracts both DNA and RNA from plant leaf tissues by simple compression and retraction, which reduces nucleic acid extraction time from hours of a conventional extraction method to less than a minute. The MN-extracted DNA/RNA is purification-free and directly applicable to the nucleic acid amplification assays such as PCR and LAMP (*ACS Nano* 2019, 13, 6540–6549). On the other side, the smartphone device comprises a sample cartridge to run the LAMP assay using MN-extracted sample and a light-emitting diode (LED) based fluorescent imaging system to quantify the LAMP assay signal. This integrated MN-smartphone nucleic acid amplification platform detects *Phytophthora infestans* DNA and tomato spotted wilt virus (TSWV) RNA from laboratory-inoculated and field-collected tomato leaves within 30 mintures. Moreover, this platform can simultaneously detect multi-infections through a single extraction and amplification on the smartphone without benchtop equipment.

634909 (Invited Talk) Development of Influenza COVID Tests with Nowdiagnostics

Abstract:

In August 2019 I began a sabbatical at NOWDiagnostics; their mission is to develop simple, accurate, and easy to administer diagnostic tests that yield results in minutes. My research initially focused on using their existing ADEXUSDx platform to develop an influenza test. I will briefly discuss our efforts on the influenza test development. As SARS-CoV-2 began to spread across the United States, it became clear that rapid and reliable tests for antigen and antibody were needed. In response to the pandemic, we quickly shifted focus to the development of antibody serology tests, antibody saliva tests, and antigen saliva tests using the existing AdexusDx platform. In this talk, I will discuss some of the obstacles we faced along with the successes.

628306 (Invited Talk) Sweatsenser Dx: An Enabling Wearable Technology for Non-Invasive Assessment Wellness Vs Illness States

Abstract:

The goal of Enlisense's SWEATSENSER Dx technology is to demonstrate the feasibility to use develop clinically relevant wearable technology to spot illness the same way many of millions of human beings track our exercise habits and sleep patterns. The core premise of the technology is rooted in proving out the hypothesis that physiological changes precede the development of symptoms for a variety of illnesses, and being able to detect these changes early may bring to the subject's attention much earlier when diseases are more readily treatable and potentially curable

In order to establish the proof-of-clinical feasibility of this technology. The SWEATSENSER has been worn by a volunteer cohort and multiple measurements have been collected in passive human sweat to tease out the body's inflammatory trending profile towards elucidating the human body's physiological response towards assessing the WELLNESS vs ILNESS state. The SWEATSENSER device captures parameters such as like heart rate, skin temperature in conjunction with biomarkers whose temporal dynamics can be mapped to a human being's physiological state. .

After obtaining baseline idea of normal readings for each person the SWEATSENSER Dx looks for deviations from these typical patterns to see whether changes might be tied to new environmental conditions, illness, or other factors that can impact health.

The goal of the SWETSENSER Dx is to create a health dashboard that does for people what dashboards already do for automobiles, Using the SWEATSENSER Dx platform we seek to demonstrate a health dashboard whereby the platform relays temporally dynamic information to the smartphone. Alerts will go off with elevated levels of pro and anti-inflammatory cytokines in conjunction with other non sweat metrics. The vision of this platform is that abnormalities to a known healthy baseline will be detected - these will enable early detection of disease, perhaps even before symptoms occur.

The overarching goal of SWEATSENSER Dx is to deliver a clinical grade technology that non-invasively monitors humans and deliver clinical-quality data to health care providers to preemptively monitor and alert them to anomalies *before* they get sick.

AUTHOR INDEX